APR 2012

D1715669

ENVIRONMENTAL SCIENCE, ENGINEERING AND TECHNOLOGY

HYDRAULIC FRACTURING AND NATURAL GAS DRILLING: QUESTIONS AND CONCERNS

ENVIRONMENTAL SCIENCE, ENGINEERING AND TECHNOLOGY

Additional books in this series can be found on Nova's website under the Series tab.

Additional E-books in this series can be found on Nova's website under the E-books tab.

ENERGY SCIENCE, ENGINEERING AND TECHNOLOGY

Additional books in this series can be found on Nova's website under the Series tab.

Additional E-books in this series can be found on Nova's website under the E-books tab.

ENVIRONMENTAL SCIENCE, ENGINEERING AND TECHNOLOGY

HYDRAULIC FRACTURING AND NATURAL GAS DRILLING: QUESTIONS AND CONCERNS

AARIK SCHULTZ
EDITOR

Nova Science Publishers, Inc.
New York

Copyright © 2012 by Nova Science Publishers, Inc.

All rights reserved. No part of this book may be reproduced, stored in a retrieval system or transmitted in any form or by any means: electronic, electrostatic, magnetic, tape, mechanical photocopying, recording or otherwise without the written permission of the Publisher.

For permission to use material from this book please contact us:
Telephone 631-231-7269; Fax 631-231-8175
Web Site: http://www.novapublishers.com

NOTICE TO THE READER

The Publisher has taken reasonable care in the preparation of this book, but makes no expressed or implied warranty of any kind and assumes no responsibility for any errors or omissions. No liability is assumed for incidental or consequential damages in connection with or arising out of information contained in this book. The Publisher shall not be liable for any special, consequential, or exemplary damages resulting, in whole or in part, from the readers' use of, or reliance upon, this material. Any parts of this book based on government reports are so indicated and copyright is claimed for those parts to the extent applicable to compilations of such works.

Independent verification should be sought for any data, advice or recommendations contained in this book. In addition, no responsibility is assumed by the publisher for any injury and/or damage to persons or property arising from any methods, products, instructions, ideas or otherwise contained in this publication.

This publication is designed to provide accurate and authoritative information with regard to the subject matter covered herein. It is sold with the clear understanding that the Publisher is not engaged in rendering legal or any other professional services. If legal or any other expert assistance is required, the services of a competent person should be sought. FROM A DECLARATION OF PARTICIPANTS JOINTLY ADOPTED BY A COMMITTEE OF THE AMERICAN BAR ASSOCIATION AND A COMMITTEE OF PUBLISHERS.

Additional color graphics may be available in the e-book version of this book.

Library of Congress Cataloging-in-Publication Data

Hydraulic fracturing and natural gas drilling : questions and concerns / editor, Aarik Schultz.
  p. cm.
 Includes bibliographical references and index.
  ISBN 978-1-61470-180-4 (hardcover : alk. paper) 1. Gas well drilling--Environmental aspects. 2. Gas well drilling--Government policy--United States. 3. Hydraulic fracturing--Environmental aspects. 4. Hydraulic fracturing--Government policy--United States. I. Schultz, Aarik.
 TD195.G3H93 2011
 622'.3381--dc23
                                                                                            2011022201

Published by Nova Science Publishers, Inc. † New York

CONTENTS

Preface		vii
Chapter 1	Hydraulic Fracturing and Safe Drinking Water Act Issues *Mary Tiemann and Adam Vann*	1
Chapter 2	Chemicals Used in Hydraulic Fracturing *United States House of Representatives Committee on Energy and Commerce Minority Staff*	49
Chapter 3	Statement of Barbara Boxer, before the Subcommittee on Water and Wildlife, Hearing on "Natural Gas Drilling: Public Health and Environmental Impacts"	89
Chapter 4	Statement of Benjamin L. Cardin, before the Subcommittee on Water and Wildlife, Hearing on "Natural Gas Drilling: Public Health and Environmental Impacts"	93
Chapter 5	Statement of James M. Inhofe, before the Subcommittee on Water and Wildlife, Hearing on "Natural Gas Drilling: Public Health and Environmental Impacts"	97

Chapter 6	Testimony of Bob Perciasepe, Deputy Administrator, U.S. Environmental Protection Agency, before the Subcommittee on Water and Wildlife	101
Chapter 7	Written Testimony of Conrad Daniel Volz, Graduate School of Public Health, University of Pittsburgh, before the Subcommittee on Water and Wildlife, Hearing on "Natural Gas Drilling: Public Health and Environmental Impacts"	107
Chapter 8	Testimony of John W. Ubinger, Jr., Senior Vice President, Pennsylvania Environmental Council, before the Committee on Environment and Public Works, Hearing on "Marcellus Shale Development in Pennsylvania"	121
Chapter 9	Testimony of Robert M. Summers, PhD., Acting Secretary of the Maryland Department of the Environment, before the Subcommittee on Water and Wildlife, Hearing on "Hydraulic Fracturing in the Marcellus Shale and Water Quality"	133
Chapter 10	Testimony of the Honorable Jeff Cloud, Oklahoma Corporation Commission, Vice Chairman, before the Subcommittee on Water and Wildlife, Hearing on "Natural Gas Drilling: Public Health and Environmental Impacts"	139
Chapter 11	Testimony of David Neslin, Director, Colorado Oil and Gas Conservation Commission, before the Subcommittee on Water and Wildlife, Hearing on "Natural Gas Drilling: Public Health and Environmental Impacts"	143
Index		149

PREFACE

Hydraulic fracturing is a technique developed initially to stimulate oil production from wells in declining oil reservoirs. More recently, it has been used to initiate oil and gas production in unconventional reservoirs where these resources were previously inaccessible. This process is now used in more than 90% of new oil and gas production wells. Hydraulic fracturing is done after a well is drilled and involves injecting large volumes of water, sand and specialized chemicals under enough pressure to fracture the formations holding the oil or gas. The sand holds the fractures open to allow the oil or gas to flow freely out of the formation and into a production well. This book explores hydraulic fracturing and safe drinking water act issues as well as the public health and environmental impacts of natural gas drilling.

Chapter 1- Hydraulic fracturing is a technique developed initially to stimulate oil production from wells in declining oil reservoirs. More recently, it has been used to initiate oil and gas production in unconventional (i.e., low-permeability) reservoirs where these resources were previously inaccessible. This process now is used in more than 90% of new oil and gas production wells. Hydraulic fracturing is done after a well is drilled and involves injecting large volumes of water, sand (or other propping agent), and specialized chemicals under enough pressure to fracture the formations holding the oil or gas. The sand or other proppant holds the fractures open to allow the oil or gas to flow freely out of the formation and into a production well.

Chapter 2- Hydraulic fracturing has helped to expand natural gas production in the United States, unlocking large natural gas supplies in shale and other unconventional formations across the country. As a result of hydraulic fracturing and advances in horizontal drilling technology, natural gas production in 2010 reached the highest level in decades. According to new estimates by the Energy Information Administration (EIA), the United States possesses natural gas resources sufficient to supply the United States for approximately 110 years.

Chapter 3- Recent advancements in horizontal drilling and hydraulic fracturing have led to a significant expansion in proven U.S. natural gas reserves. Natural gas resources are now recoverable that were considered uneconomical even a few years ago.

The discovery of new resources creates an opportunity for increased production of a cleaner, domestically-produced fuel. While this could have benefits for our nation's economy and energy independence, it is critical to ensure that exploration for natural gas is done safely and responsibly.

Chapter 4- We have enormous reserves that can help America meet its energy needs and do so in a way that produces far less pollution than coal, helps the United States on its path to energy independence, and improves national security.

High volume, horizontal hydraulic fracturing, or "fracking," is now being used to extract natural gas from shale formations in thousands of new wells. In Pennsylvania more than 2,700 Marcellus wells were drilled from 2006 to March 10th of this year.

Chapter 5- Now let me show you why this is the case. This chart illustrates a cross section of a typical well drilled in the Marcellus shale in southwest Pennsylvania. Do you see the small blue line at the top of the chart? That illustrates the ground water aquifer. In between that groundwater aquifer and the Marcellus shale are dozens of layers of solid rock – *more than a mile of it*. Let me say that again: there is more than a mile that separates the groundwater aquifer and the well.

Chapter 6- Natural gas can enhance our domestic energy options, reduce our dependence on foreign supplies, and serve as a bridge fuel to renewable energy sources. If produced responsibly, natural gas has the potential to improve air quality, stabilize energy prices, and provide greater certainty about future energy reserves.

Chapter 7- Unconventional gas extraction in deep shale deposits presents considerable risks to public health and safety as well as to environmental resources, particularly water quality and aquatic organisms. My testimony today will cover three critical public health and environmental policy areas related to unconventional natural gas production.

Chapter 8- The Marcellus Shale is one of the largest unconventional on-shore gas deposits in the world. Estimated at between 250-500 trillion cubic feet of gas deep underground, the Marcellus Shale represents a natural gas supply that could meet America's energy needs for the next 50-80 years or more.

Chapter 9- The Marcellus Shale formation underlies Garrett County and part of Allegany County in the far western portion of Maryland. In these two counties, gas companies have leased the gas rights on more than 100,000 acres. The Maryland Department of the Environment issues permits for oil and gas wells, and we received our first permit application for drilling and hydraulic fracturing ("fracking") in the Marcellus Shale in 2009. No permits have yet been issued. We currently have applications pending from two companies for a total of 5 wells. We are mindful of the tremendous benefits that could accrue to the environment and the economy by exploring and exploiting our gas reserves, but we are equally alert to the risks of adverse public health and environmental effects. Our paramount concern is protecting our ground and surface waters.

Chapter 10- Oklahoma's first commercial oil well was drilled in 1897, which was 10 years before Oklahoma officially became a state in 1907. Since then, oil and natural gas production has expanded into almost part of the state.

The Oklahoma Corporation Commission (OCC) was first given responsibility for regulation of oil and gas production in Oklahoma in 1914. Currently the Commission has exclusive state jurisdiction over all oil and gas industry activity in Oklahoma, including oversight and enforcement of rules aimed at pollution prevention and abatement and protecting the state's precious water supplies.

Chapter 11- Colorado has a long and proud history of oil and gas development, with our first oil well dating back to 1862. As of 2009, we ranked fifth in natural gas production and tenth in oil production. Our diverse hydrocarbon resources encompass a variety of shale, tight sand, coal bed methane, and other formations that span the state. At the same time, we have a thriving resort and tourist economy, and our rugged mountains, clear streams, and abundant wildlife are an essential part of our heritage.

In: Hydraulic Fracturing and Natural Gas Drilling ISBN: 978-1-61470-180-4
Editor: Aarik Schultz © 2012 Nova Science Publishers, Inc.

Chapter 1

HYDRAULIC FRACTURING AND SAFE DRINKING WATER ACT ISSUES[*]

Mary Tiemann and Adam Vann

SUMMARY

Hydraulic fracturing is a technique developed initially to stimulate oil production from wells in declining oil reservoirs. More recently, it has been used to initiate oil and gas production in unconventional (i.e., low-permeability) reservoirs where these resources were previously inaccessible. This process now is used in more than 90% of new oil and gas production wells. Hydraulic fracturing is done after a well is drilled and involves injecting large volumes of water, sand (or other propping agent), and specialized chemicals under enough pressure to fracture the formations holding the oil or gas. The sand or other proppant holds the fractures open to allow the oil or gas to flow freely out of the formation and into a production well.

Its application, along with horizontal drilling, for production of natural gas (methane) from coal beds, tight gas sands, and, more recently, from unconventional shale formations, has resulted in the marked expansion of estimated U.S. natural gas reserves in recent years. Similarly, hydraulic fracturing is enabling the development of unconventional domestic oil resources, such as the Bakken Formation in

---

[*] This is an edited, reformatted and augmented version of a Congressional Research Service publication, CRS Report for Congress R41760, from www.crs.gov, dated April 7, 2011.

North Dakota and Montana. However, the rapidly increasing and geographically expanding use of fracturing, along with a growing number of citizen complaints and state investigations of well water contamination attributed to this practice, has led to calls for greater state and/or federal environmental regulation and oversight of this activity.

Historically, the Environmental Protection Agency (EPA) had not regulated the underground injection of fluids for hydraulic fracturing of oil or gas production wells. In 1997, the U.S. Court of Appeals for the 11th Circuit ruled that fracturing for coalbed methane (CBM) production in Alabama constituted underground injection and must be regulated under the Safe Drinking Water Act (SDWA). This ruling led EPA to study the risk that hydraulic fracturing for CBM production might pose to drinking water sources. In 2004, EPA reported that the risk was small, except where diesel was used, and that regulation was not needed. However, to address regulatory uncertainty the ruling created, the Energy Policy Act of 2005 (EPAct 2005) revised the SDWA term "underground injection" to explicitly exclude the injection of fluids and propping agents (except diesel fuel) used for hydraulic fracturing purposes. Consequently, EPA currently lacks authority under the SDWA to regulate hydraulic fracturing, except where diesel fuel is used. However, as the use of this process has grown, some in Congress would like to revisit this statutory exclusion.

In the 112th Congress, H.R. 1084 and S. 587, the Fracturing Responsibility and Awareness of Chemicals Act (FRAC Act), have been introduced. The legislation would repeal the exemption for hydraulic fracturing operations that was established in EPAct 2005, and would amend the term "underground injection" to include explicitly the injection of fluids used in hydraulic fracturing operations related to oil and gas production, thus authorizing EPA to regulate this process under the SDWA. The bills also would require disclosure of the chemicals used in the fracturing process. EPA's FY2010 appropriations act directed EPA to study the relationship between hydraulic fracturing and drinking water. The interim report, expected in 2012, may help inform Congress on whether federal action is needed. Meanwhile, various states are reviewing, and some have revised, their oil and gas rules to address the increased use of hydraulic fracturing.

This report reviews past and proposed treatment of hydraulic fracturing under the SDWA, the principal federal statute for regulating the underground injection of fluids to protect groundwater sources of drinking water. It reviews current SDWA provisions for regulating underground injection activities, and discusses some possible implications of, and issues associated with, enactment of legislation authorizing EPA to regulate hydraulic fracturing under this statute.

INTRODUCTION

Background—Hydraulic Fracturing in Oil and Gas Production

The process of hydraulic fracturing was developed initially in the 1940s to stimulate production from oil reservoirs with declining productivity. More recently, this practice has been used to initiate oil and gas production in unconventional (low-permeability) oil and gas formations.[1] Its application—in combination with technological breakthroughs, such as horizontal drilling—in the production of natural gas from coal beds, tight gas sands,[2] and unconventional shale formations has resulted in the marked expansion of estimated U.S. natural gas reserves in recent years. Similarly, hydraulic fracturing is enabling the development of unconventional domestic oil resources, such as the Bakken Formation in North Dakota and Montana. However, the rapidly increasing and geographically expanding use of fracturing, along with a growing number of complaints of well water contamination and other water quality problems attributed to this practice, has led to calls for greater state and/or federal oversight of this activity.

Hydraulic fracturing involves injecting into production wells large volumes of water, sand or other proppant,[3] and specialized chemicals under enough pressure to fracture low-permeability geologic formations containing oil and/or natural gas.[4] The sand or other proppant holds the new fractures open to allow the oil or gas to flow freely out of the formation and into a production well. Fracturing fluid and water remaining in the fracture zone can inhibit oil and gas production, and must be pumped back to the surface. The fracturing fluid—"flowback"—along with any naturally occurring formation water pumped to the surface, together called produced water, typically has been disposed through deep well injection or treated before disposal into surface waters.[5] According to industry estimates for various geographic areas, the volume of flowback water can range from less than 30% to more than 70% of the original fracture fluid volume.[6]

The use of hydraulic fracturing continues to increase significantly, as more easily accessible oil and gas reservoirs have declined and companies move to develop unconventional oil and gas formations. Hydraulic fracturing is used for oil and/or gas production in all 33 U.S. states where oil and natural gas production takes place. According to industry estimates, hydraulic fracturing has been applied to more than 1 million wells nationwide.[7]

The frequency of its use expanded markedly in the 1980s and 1990s with its application to coalbed methane (CBM) development. CBM production through wells began in the 1970s as a safety measure in coal mines to reduce the explosion hazard posed by methane. In 1984, fewer than 100 coalbed wells existed in the United States.[8] In the 1980s, demand for natural gas, new fracturing technologies, and federal tax credits for nonconventional fuels production led to dramatic growth in the CBM development industry. By 1990, nearly 8,000 coalbed wells had been drilled nationwide. In 2008, the Environmental Protection Agency (EPA) identified 56,000 CBM wells managed by operators in 692 different projects.[9]

Other unconventional gas resource formations relying on hydraulic fracturing include tight sands gas and shale gas. The Department of Energy's (DOE's) Energy Information Agency (EIA) reports that natural gas from tight sand formations is the largest source of unconventional production, while production from shale formations is the fastest growing source.[10] Figure 1 illustrates different types of natural gas reservoirs.

The number of onshore gas wells in the United States increased from approximately 260,000 wells in 1989 to 493,100 wells in 2009.[11] According to the Independent Petroleum Association of America (IPAA), more than 90% of new natural gas wells in the United States rely on hydraulic fracturing, and together they have accounted for the production of more than 600 trillion cubic feet of gas. Similarly, fracturing is increasingly applied to U.S. oil production, and more than 7 billion barrels of oil have been produced using this process.

As noted, it is the combination of hydraulic fracturing and directional drilling that is allowing the economic development of unconventional oil and gas resources. Improvements in technology also have led to increased use of horizontal drilling in developing unconventional gas formations. Currently, shale gas production involves drilling a well vertically and then drilling horizontally out from the wellbore. Because of the low permeability of shales, more wells must be drilled into a reservoir than more permeable, conventional reservoirs. A benefit of horizontal drilling through a producing shale layer is that one well pad that utilizes horizontal well drilling can replace numerous individual well pads and reduce the surface density of wells in an area. Six to eight horizontal wells, and potentially more, can be drilled from a single well pad and access the same reservoir. According to a report prepared for DOE,

> the spacing interval for vertical wells in the gas shale plays averages 40 acres per well for initial development. The spacing interval for horizontal wells is likely to be approximately 160 acres per well. Therefore, a 640-acre section of land could be developed with a total of 16 vertical wells, each on its own individual well pad, or by as few as 4 horizontal wells all drilled from a single multi-well drilling pad.[12]

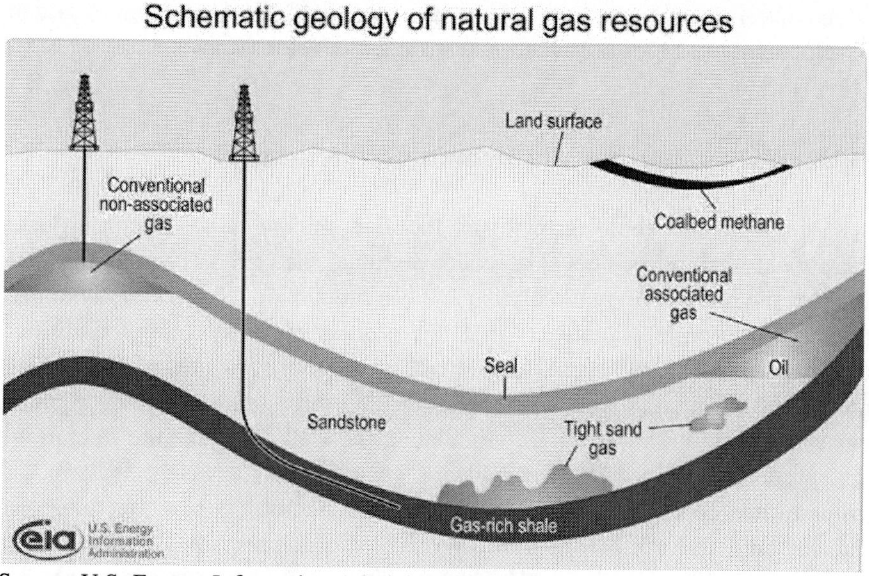

Schematic geology of natural gas resources

Source: U.S. Energy Information Administration, Independent Statistics and Analysis, October 2008. Available at http://www.eia.gov/oil

Notes: The diagram shows schematically the geologic nature of most major U.S. sources of natural gas:

- Gas-rich shale is the source rock for many natural gas resources, but, until [recently], has not been a focus for production. Horizontal drilling and hydraulic fracturing have made shale gas an economically viable alternative to conventional gas resources.
- Conventional gas accumulations occur when gas migrates from gas rich shale into an overlying sandstone formation, and then becomes trapped by an overlying impermeable formation, called the seal. Associated gas accumulates in conjunction with oil, while non-associated gas does not accumulate with oil.
- Tight sand gas accumulations occur in a variety of geologic settings where gas migrates from a source rock into a sandstone formation, but is limited in its ability to migrate upward due to reduced permeability in the sandstone.
- Coalbed methane does not migrate from shale, but is generated during the transformation of organic material to coal.

Figure 1. Geologic Nature of Major Sources of Natural Gas in the United States.

A single gas production well may be fractured multiple times, using from 500,000 gallons to more than 6 million gallons of water, with compounds and proppants of various amounts added to the water.[13] Slickwater fracturing, which involves adding chemicals to increase fluid flow, is a more recent development used for unconventional shale gas development.[14]

Hydraulic Fracturing and Drinking Water Issues

Although the rapid growth in the use of hydraulic fracturing and directional drilling to develop unconventional natural gas resources has enabled the industry to markedly expand gas production, concern has emerged regarding the potential impacts that this process may have on groundwater quality and specifically on private wells and public water supplies. The process of developing a shale gas well (drilling through an overlying aquifer, completing and casing the well, stimulating the well via hydraulic fracturing, and producing the gas) is an issue of concern for increasing the risk of groundwater contamination. During hydraulic fracturing, new fractures are induced into the shale formation, or existing fractures are lengthened. As exploration and production activities have increased and expanded into more populated areas, so has concern that the hydraulic fracturing process might introduce chemicals, methane, and other contaminants into aquifers.

Another concern involves the potential contamination of drinking water wells from surface activities. A water well that is not constructed and cased properly might be at risk if contaminated water flows from the land surface and enters the water well, possibly compromising the quality of drinking water in the well and even the aquifer itself. In such instances, and particularly where natural gas drilling and stimulation activities are nearby, leaky surface impoundments, accidental spills, or careless surface disposal of drilling fluids at the natural gas production site could increase the risk of contaminating the nearby water well.

Other water quality concerns involve the management (storage, treatment and disposal) of water produced in the fracturing process. Broader environmental issues associated with the more concentrated and geographically expanding development of unconventional gas resources include water withdrawals from streams, lakes and aquifers; potential air quality impacts; and land use changes (including those related to the development of access roads, pipelines and drill pads). Although such issues can be significant for state regulators, gas developers, local communities and landowners; they are not addressed in this report.[15]

Public complaints of impacts to well water have increased as gas development has intensified. In 2009, the Ground Water Protection Council

(GWPC)[16] reported that several citizen complaints of well contamination attributed to hydraulic fracturing appeared to be related to hydraulic fracturing of CBM zones that were in relatively close proximity to underground sources of drinking water.[17] Additional contamination incidents in other gas producing areas have been reported.[18] In Pennsylvania, for one example, regulators confirmed that methane had migrated to water wells from drilling sites and issued notices of violations to a drilling company for, among other things, "failure to prevent gas from entering fresh groundwater."[19] Other incidents are under investigation.

Additionally, EPA Region 8 has investigated the potential role of hydraulic fracturing in the contamination of a cluster of water wells in the Pavillion, WY, area. Using its authority under the Comprehensive Environmental Response, Cleanup, and Liability Act (CERCLA, commonly known as Superfund),[20] EPA began testing water wells after residents contacted EPA in 2008 to report changes in the quality and quantity of water following nearby gas development. The agency confirmed the presence of a compound known to be used in hydraulic fracturing in several wells, and is continuing to analyze the results of its investigations.[21]

In many cases, the source or cause of well-water contamination remains undetermined. Identifying the cause of contamination can be difficult for various reasons, including the complexity of hydrogeologic processes and investigations, a lack of baseline testing of nearby water wells prior to drilling and fracturing, as well as the confidential business information status typically given to fracturing compounds across the states. In other cases, contamination incidents have been attributed to poor well construction or surface activities, rather than the specific hydraulic fracturing process. Responding to a recent survey, major oil and gas producing states asserted that the hydraulic fracturing process has not been linked directly to groundwater contamination. However, contamination incidents attributed to poor well construction have raised concerns regarding the adequacy and/or enforcement of state well construction regulations (covering, for example, cementing, casing, and backflow prevention) for purposes of managing oil and gas development that is increasingly dependent on fracturing.

A key barrier to better understanding groundwater contamination risks that may be associated with hydraulic fracturing is the lack of scientific studies to assess the practice and related complaints. A further issue is that EPA developed the underground injection control (UIC) program primarily to regulate wells that received fluids injected for the long term or for enhanced recovery operations, but excluded oil and gas *production* wells. This distinction could raise regulatory challenges and the possibility that the agency may need to develop an essentially new framework to address hydraulic fracturing of production wells.

Such information gaps and regulatory issues contribute to uncertainty over a possible legislative or regulatory framework that might be developed for hydraulic fracturing activities, as well as the potential costs and benefits associated with any measures. The sheer number of wells that rely on fracturing, typically multiple times, suggests that significant new resources could be required by state and federal regulators to implement and enforce any new EPA requirements on top of existing state requirements. Similarly, oil and gas industry representatives have expressed concern that well owners and operators might experience impacts, such as higher operation and compliance costs, and delays in permitting and disruption of operations, particularly early on, as any regulatory requirements are put into place and state programs are revised and then reviewed and approved by EPA (and as EPA develops regulations and implements requirements directly for nonprimacy states). States would also have to integrate any new requirements and programs with their existing oil and gas regulatory programs. The adjustments would vary among the states, reflecting different state rules and regulatory structures.

Experience regulating hydraulic fracturing in Alabama and regulatory developments in several states (including Colorado, New York, Pennsylvania, and Wyoming) to address the growing reliance on hydraulic fracturing may add insight to the possible implications of proposed federal legislation and any subsequent regulations. Additionally, Congress has directed EPA to conduct a study on the relationship between hydraulic fracturing and drinking water.[22] The Agency expects to report on the interim research results in 2012, and issue a follow-up report in 2014.

THE SAFE DRINKING WATER ACT AND THE FEDERAL ROLE IN REGULATION OF UNDERGROUND INJECTION

Review of Relevant SDWA UIC Provisions

To properly evaluate studies and any new federal or state action, it is important to understand the existing statutory and regulatory framework. Most public water systems and nearly all rural residents rely on groundwater as a source of drinking water. Because of the nationwide importance of underground sources of drinking water, Congress included groundwater protection provisions in the 1974 Safe Drinking Water Act (SDWA). The SDWA, among other things, directs the EPA to regulate the underground injection of fluids (including solids, liquids, and gases) to protect underground sources of drinking water.[23]

Part C of the SDWA establishes the national regulatory program for the protection of underground sources of drinking water, including the oversight and limitation of underground injections that could affect aquifers through the establishment of underground injection control (UIC) regulations. Key UIC requirements and exceptions contained in SDWA, Part C, include:

- Section 1421 of the SDWA directs the EPA Administrator to promulgate regulations for state UIC programs, and mandates that the EPA regulations "contain minimum requirements for programs to prevent underground injection that endangers drinking water sources." Section 1421(b)(2) specifies that EPA:

    may not prescribe requirements for state UIC programs which interfere with or impede—(A) the underground injection of brine or other fluids which are brought to the surface in connection with oil or natural gas production or natural gas storage operations, or (B) any underground injection for the secondary or tertiary recovery of oil or natural gas, *unless such requirements are essential to assure that underground sources of drinking water will not be endangered by such injection.*[24] [emphasis added]

- Section 1421(d), as amended by Energy Policy Act of 2005 (EPAct 2005),[25] specifies that the term "underground injection" as it is used in the SDWA, means the subsurface emplacement of fluids by well injection, and specifically excludes the underground injection of fluids or propping agents associated with hydraulic fracturing operations related to oil, gas, or geothermal production activities.[26] The use of diesel fuels in hydraulic fracturing, however, forfeits eligibility for this exclusion from the definition of "underground injection."[27]

- Section 1422 authorizes EPA to delegate primary enforcement authority (primacy) for UIC programs to the states, provided that the state program meets EPA requirements promulgated under Section 1421 and prohibits any underground injection that is not authorized by a state permit or rule.[28] If a state's UIC program plan is not approved, or the state has chosen not to assume program responsibility, then EPA must implement the UIC program in that state.

- Section 1425 authorizes EPA to approve the portion of a state's UIC program that relates to "any underground injection for the secondary or tertiary recovery of oil or natural gas" if the state program meets certain requirements of § 1421 and represents an effective program to prevent underground injection which endangers drinking water sources.[29] Under this provision, states may demonstrate to EPA that their existing programs for oil and gas injection wells are effective in preventing endangerment of underground sources of drinking water. This provides states with an alternative to meeting the specific requirements contained in EPA regulations promulgated under Section 1421.[30] (See discussion on p. 11.)
- Section 1423 authorizes EPA enforcement actions for UIC regulatory violations.
- Section 1431 applies broadly to the SDWA and grants the EPA Administrator emergency powers to issue orders and commence civil actions to protect public water systems or underground sources of drinking water.[31]
- Section 1449, another broadly applicable SDWA provision, authorizes citizen civil actions against persons allegedly in violation of the act's enforceable requirements, or against EPA for allegedly failing to perform a duty. State-administered oil and gas programs may not have such provisions, so this could represent an expansion in the ability of citizens to challenge administration of statutes and regulations related to hydraulic fracturing and drinking water, were the hydraulic fracturing exemption provision to be repealed.

The "Endangerment" Standard

As noted, the SDWA states that UIC regulations must "contain minimum requirements for effective programs to prevent underground injection which endangers drinking water sources."[32] Known as the "endangerment standard," this statutory standard is a major driving force in EPA regulation of underground injection.

The endangerment language focuses on protecting groundwater that is used or may be used to supply public water systems. This focus parallels the general scope of the statute, which addresses the quality of water provided by public water systems and does not address private, residential wells. The endangerment language has raised questions as to whether EPA regulations

Hydraulic Fracturing and Safe Drinking Water Act Issues

can reach underground injection activities to protect groundwater that is not used by public water systems.

Defining "Underground Source of Drinking Water"

The SDWA directs EPA to protect against endangerment of an "underground source of drinking water" (USDW). The statute defines a USDW to mean an aquifer or part of an aquifer that either:

- supplies a public water system, or
- contains a sufficient quantity of ground water to supply a public water system;[33] and
    - currently supplies drinking water for human consumption; or
    - contains fewer than 10,000 milligrams per liter (mg/L) total dissolved solids; and
- is not an "exempted aquifer."[34]

In a 2004 report on hydraulic fracturing of coalbed methane reservoirs, the agency further noted that the "EPA also assumes that all aquifers contain sufficient quantity of groundwater to supply a public water system, unless proven otherwise through empirical data."[35] However, because these expanded agency characterizations of what constitutes a USDW are not included in SDWA or related regulation, and, therefore, are not binding on the agency, it is uncertain how they might be applied in future situations. Notably, the SDWA does not prohibit states from establishing requirements that are stricter than federal requirements, and many states have their own definitions and classifications for groundwater resources.

Underground Injection Control Regulatory Program Overview

To implement the UIC program as mandated by the provisions of the SDWA described above, EPA has established six classes of underground injection wells based on categories of materials that are injected by each class into the ground. In addition to the similarity of fluids injected in each class of wells, each class shares similar construction, injection depth, design and operating techniques. The wells within a class are required to meet a set of appropriate performance criteria for protecting underground sources of drinking water (USDW). The six well categories are briefly described below, including the estimated number of wells nationwide.[36]

- Class I wells inject hazardous wastes, industrial non-hazardous liquids, or municipal wastewater beneath the lowermost USDW. (There are 549 such wells regulated as Class I wells in the United States). The most stringent regulations apply to these wells.
- Class II wells inject brines and other and fluids associated with oil and gas production, and hydrocarbons for storage. The wells inject fluids beneath the lowermost USDW (143,951 wells). Section 1425, which allows states to apply their own regulations in lieu of EPA regulations, applies to Class II wells.
- Class III wells inject fluids associated with solution mining of minerals (e.g., salt and uranium) beneath the lowermost USDW (18,505 wells).
- Class IV wells inject hazardous or radioactive wastes into or above USDWs. These wells are banned unless authorized under a federal or state ground water remediation project (32 wells).
- Class V includes all injection wells not included in Classes I-IV, including experimental wells. Class V wells frequently inject non-hazardous fluids into or above USDWs and are typically shallow, on-site disposal systems. However, some deep Class V wells inject below USDWs (400,000-650,000).[37]
- Class VI wells: In 2010, EPA issued a rule for Class VI wells, to be used for the geologic sequestration of carbon dioxide.

The UIC regulatory program includes the following broad elements: site characterization, area of review, well construction, well operation, site monitoring, well plugging and post-injection site care, public participation, and financial responsibility. While the six classes broadly share similar regulatory requirements, those for Class I wells are the most comprehensive and stringent. Table 1 outlines the shared minimum technical requirements for Class I, II, and III wells.

Table 1. Minimum Federal Technical Requirements for Class I, II, and III Wells

Permitting Requirements Common to Class I, II, and III Wells
Demonstration that casing and cementing are adequate to prevent movement of fluid into or between USDWs. Cement bond logs are often needed to evaluate/verify the adequacy of the cementing records.
Financial assurances (bond, letter of credit, or other adequate assurance) that the owner or operator will maintain financial responsibility to properly plug and abandon the wells.
A maximum operating pressure calculated to avoid initiating and/or propagating fractures that would allow fluid movement into a USDW.
Monitoring and reporting requirements.
Requirement that all permitted (and rule authorized) wells which fail mechanical integrity be shut in immediately. A well may not resume injection until mechanical integrity has been demonstrated.
Schedule for demonstrating mechanical integrity (at least every five years for Class I nonhazardous, Class II, and Class III salt recovery wells).[a]
All permitted injection wells, which have had the tubing disturbed, must have a pressure test to demonstrate mechanical integrity.
Plans for plugging and abandonment. All Class I, II, and III wells must be plugged with cement.

Source: U.S. EPA Technical Program Overview: Underground Injection Control Regulations, EPA 816-R-02-025, December 2002, p. 65.

[a] Class I hazardous wells must demonstrate mechanical integrity once a year.

Class II Wells

Because this discussion of hydraulic fracturing is related to oil and gas production, this report focuses primarily on regulatory requirements for Class II wells rather than other categories of wells in EPA's UIC program. If authorized or mandated to regulate hydraulic fracturing broadly under SDWA, EPA might regulate hydraulic fracturing as a Class II activity, which would parallel its approach for regulating the injection of diesel for fracturing purposes. However, it is possible that EPA could classify oil and gas production wells that are hydraulically fractured under a different class, or develop an entirely new regulatory structure or subclass of wells.[38]

A Class II well may be used to dispose of brines (salt water) and other fluids associated with oil and gas production or storage, to store natural gas, or to inject fluids for enhanced oil and gas recovery. EPA estimates that some 80% of Class II wells are enhanced recovery (ER) wells. These wells inject brine, water, stream polymers, or carbon dioxide primarily into oil-bearing formations (also called secondary or tertiary recovery). Enhanced recovery wells are separate from, and typically surrounded by, production wells.[39] Table 2 outlines basic requirements for Class II wells.

Table 2. Minimum EPA Regulatory Requirements for Class II Wells

Requirement	Explanation
Permit Required	Yes, except for existing Enhanced Oil Recovery (EOR) wells authorized by rule
Life of Permit	Specific period, may be for life of well
Area of Review	New wells—¼ mile fixed radius or radius of endangerment
Mechanical Integrity Test (MIT) Required	Internal MIT: prior to operation, and pressure test or alternative at least once every five years for internal well integrity. External MIT: cement records may be used in lieu of logs.
Other Tests	Annual fluid chemistry and other tests as needed/required by permit
Monitoring	Injection pressure, flow rate and cumulative volume, observed weekly for disposal and monthly for enhanced recovery
Reporting	Annual

Source: U.S. Environmental Protection Agency, Technical Program Overview: Underground Injection Control Regulations, EPA 816-R-02-025, July 2001, p. 11, 67, and Appendix E.

State Primacy for UIC Program Administration

Thirty-three states have assumed primacy for the UIC program, EPA has lead implementation authority in 10 states, and authority is shared in the remaining states. EPA directly implements the entire UIC program in several oil and gas producing states, including Kentucky, Michigan, New York, Pennsylvania, and Virginia.[40] Figure 2 identifies state primacy status for the UIC program.

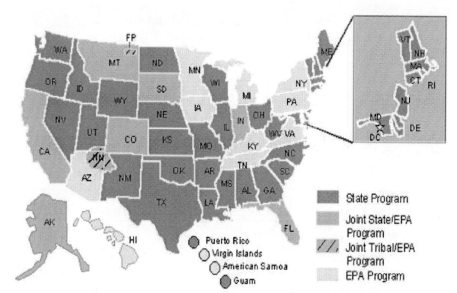

* The Fort Peck (FP) Tribes and the Navajo Nation (NN) are currently the only Tribes with UIC Primacy.
Source: U.S. Environmental Protection Agency, available at http://www.epa.gov/safewater/uic/primacy

Figure 2. Primacy Status for EPA's UIC Program.

As noted, for oil and gas related injection operations, under Section 1425, a state may be delegated primary enforcement authority without meeting EPA regulatory requirements for state UIC programs promulgated under Section 1421, provided the state demonstrates that it has an effective program that prevents underground injection that endangers drinking water sources. EPA has issued guidance for approval of state programs under Section 1425.[41] If directed by Congress to regulate hydraulic fracturing as underground injection, this regulatory approach could give states significant flexibility and thus might reduce potential regulatory costs, redundancy, and other possible impacts to the industry and the states.[42]

Most oil and gas producing states exercise primary enforcement authority for injection wells associated with oil and gas production (Class II wells) under SDWA Section 1425. Among these states, Alaska, California, Colorado, Indiana, Montana, and South Dakota have received primacy only for Class II wells, while EPA administers the remainder of the UIC program (Class I, III, IV, and V wells) for these states. Table 3 lists states that regulate Class II wells under Section 1425.

**Table 3. States Regulating Oil and Gas
(Class II) UIC Wells Under SDWA § 1425**

Alabama	Louisiana	Oklahoma
Alaska	Mississippi	Oregon
Arkansas	Missouri	South Dakota
California	Montana	Texas
Colorado	Nebraska	Utah
Illinois	New Mexico	West Virginia
Indiana	North Dakota	Wyoming
Kansas	Ohio	

Source: Adapted from information provided by U.S. Environmental Protection Agency.

Note: With primacy granted under § 1425, states regulate Class II wells using their own program requirements rather than following EPA regulations.

THE DEBATE OVER REGULATION OF HYDRAULIC FRACTURING UNDER THE SDWA

From the date of the enactment of the SDWA in 1974 until the late 1990s, hydraulic fracturing was not regulated under the act by EPA or the states tasked with administration of the SDWA. However, in the last 15 years a number of developments have called into question the extent to which hydraulic fracturing should be considered an "underground injection" to be regulated under the SDWA. One trigger for this debate was a challenge to the Alabama UIC program brought by the Legal Environmental Assistance Foundation (LEAF).

The LEAF Challenge to the Alabama UIC Program and the EPA's Interpretation of the SDWA

In 1994, LEAF petitioned EPA to initiate proceedings to have the agency withdraw its approval of the Alabama UIC program because the program did not regulate hydraulic fracturing operations in the state associated with production of methane gas from coalbed formations.[43] The State of Alabama had previously been authorized by EPA to administer a UIC program pursuant to the terms of the SDWA.[44] EPA denied the LEAF petition in 1995 based on a finding that hydraulic fracturing did not fall within the definition of "underground injection" as the term was used in the SDWA and the EPA regulations promulgated under that act.[45] According to EPA, that term applied only to wells whose "principal function" was the placement of fluids underground.[46] LEAF challenged EPA's denial of its petition in the U.S. Court of Appeals for the Eleventh Circuit, arguing that EPA's interpretation of the terms in question was inconsistent with the language of the SDWA.[47]

The court rejected EPA's claim that the language of the SDWA allowed it to regulate only those wells whose "principal function" was the injection of fluids into the ground. EPA based this claim on what it perceived as "ambiguity" in the SDWA regarding the definition of "underground injection" as well as a perceived congressional intent to exclude wells with primarily non-injection functions.[48] The court held that there was no ambiguity in the SDWA's definition of "underground injection" as "the subsurface emplacement of fluids by well injection," noting that the words have a clear meaning and that:

> The process of hydraulic fracturing obviously falls within this definition, as it involves the subsurface emplacement of fluids by forcing them into cracks in the ground through a well. Nothing in the statutory definition suggests that EPA has the authority to exclude from the reach of the regulations an activity (i.e. hydraulic fracturing) which unquestionably falls within the plain meaning of the definition, on the basis that the well that is used to achieve that activity is also used—even primarily used—for another activity (i.e. methane gas production) that does not constitute underground injection.[49]

The court therefore remanded the decision to EPA for reconsideration of LEAF's petition for withdrawal of Alabama's UIC program approval.[50]

Alabama's Regulation of Hydraulic Fracturing in CBM Production

Consideration of Alabama's UIC program after the *LEAF I* decision was issued in 1997 is a helpful case study. It is useful in assessing exactly how EPA authorized a state to regulate hydraulic fracturing under the SDWA

"Class" well system, understanding the regulatory options available to EPA and the states authorized to enforce SDWA programs, and evaluating the industry impact resulting from the requirement that hydraulic fracturing be regulated under a UIC program.

Following the *LEAF I* decision and EPA's initiation of proceedings to withdraw its approval of Alabama's Class II UIC program, in 1999 Alabama submitted a revised UIC program to EPA.[51] The revised UIC program sought approval under Section 1425 of the SDWA rather than Section 1422(b). As discussed above, Section 1425 differs from Section 1422(b) in that approval under Section 1425 is based on a showing by the state that the program meets the generic requirements found in Section 1421(b)(1)(A)-(D) of the SDWA and that the program "represents an effective program (including adequate recordkeeping and reporting) to prevent underground injection which endangers drinking water sources."[52] In contrast, approval of a state program under Section 1422(b) requires a showing that the state's program satisfies the requirements of the UIC regulations promulgated by EPA.[53] In its decision on the challenge to EPA's approval of Alabama's revised UIC program, the U.S. Court of Appeals for the Eleventh Circuit observed that "the practical difference between the two statutory methods for approval is that the requirements for those programs covered under § 1425 are more flexible than the requirements for those programs covered under § 1422(b)."[54]

EPA approved Alabama's revised UIC program under Section 1425 in 2000.[55] LEAF appealed EPA's decision to the U.S. Court of Appeals for the Eleventh Circuit. LEAF made three arguments. First, LEAF claimed that EPA should not have approved state regulation of hydraulic fracturing under Section 1425 because it does not "relate to ... underground injection for the secondary or tertiary recovery of oil or natural gas," one of the requirements for approval under Section 1425.[56] The court rejected this argument, finding that the phrase "relates to" was broad and ambiguous enough to include regulation of hydraulic fracturing as being related to secondary or tertiary recovery of oil or natural gas.[57]

Second, LEAF challenged the Alabama program's regulation of hydraulic fracturing as "Class II-like" wells not subject to the same regulatory requirements as Class II well.[58] The court agreed with LEAF on this point, noting that in its decision in *LEAF I*, it had held that methane gas production wells used for hydraulic fracturing are "wells" within the meaning of the statute.[59] As a result, the court found that wells used for hydraulic fracturing must fall under one of the five classes set forth in the EPA regulations at 40 C.F.R. § 144.6.[60] Specifically, the court found that the injection of hydraulic fracturing fluids for recovery of coalbed methane "fit squarely within the definition of Class II wells," and as a result the court remanded the matter to EPA for a determination of whether Alabama's updated UIC program complied with the requirements for Class II wells.[61]

Finally, LEAF alleged that even if Alabama's revised UIC program was eligible for approval under Section 1425 of the SDWA, EPA's decision to approve it was "arbitrary and capricious" and therefore a violation of the Administrative Procedure Act.[62] The court rejected this argument.[63]

Among other provisions added in response to the Eleventh Circuit's decisions, the Alabama regulations prohibited fracturing "in a manner that would allow the movement of fluid containing any contaminant into a USDW, if the presence of the contaminant may (a) cause a violation of any applicable primary drinking water standard; or (b) otherwise adversely affect the health of persons."[64] The state regulations further required state approvals (but not permits) prior to individual fracturing jobs. Specifically, well operators were required to certify in writing, with supporting evidence, that a proposed hydraulic fracturing operation would not occur in a USDW, or that the mixture of fracturing fluids would meet EPA drinking water standards. Regulations also prohibited fracturing at depths shallower than 399 feet (most drinking water wells rely on shallow aquifers) and prohibited the use of diesel oil or fuel in any fracturing fluid mixture. The requirements regarding minimum depths and the diesel ban remain in place, but the rules no longer require that injection fluids meet drinking water standards. Instead, "each coal bed shall be hydraulically fractured so as not to cause irreparable damage to the coalbed methane (CBM) well, or to adversely impact any fresh water supply well or any fresh water resources."[65]

With hydraulic fracturing regulations in place, CBM development in Alabama continued. In 2009, a member of the State Oil and Gas Board of Alabama noted, "since Alabama adopted its hydraulic fracturing regulations, coalbed operators have submitted thousands of hydraulic fracturing proposals and engaged in thousands of hydraulic fracturing operations."[66]

The number of CBM well permits increased in the years following the adoption of revised regulations.[67] However, it is not clear whether, or by how much, the number of wells, the production costs, or the time required by operators may have been different without the revisions.[68] One of the requirements of the Alabama regulations in response to *LEAF I* was that fracturing fluids had to meet tap water standards where fracturing would occur within an underground source of drinking water. To ensure compliance, operators purchased water from municipal water supplies that were in compliance with federal drinking water standards to use for fracturing wells. Industry representatives have noted that if this approach were adopted for hydraulic fracturing nationwide, it would not only raise costs, but potentially put companies in competition with communities for drinking water supplies.

Some concern has been expressed that if Congress passed legislation requiring federal regulation of hydraulic fracturing broadly,[69] a separate permit might be required each time a well is hydraulically fractured, thus repeatedly disrupting oil and gas production activities. In Alabama, in response to *LEAF I*, the state did not require a permit for each fracturing operation, but rather had operators give notice and receive approval before fracturing. To further facilitate approvals for hydraulic fracturing, service companies identified to the state chemicals contained in various fracturing fluid mixtures that met the regulatory requirement that the mixtures not exceed federal drinking water standards. A well operator then could select from a list of pre-approved hydraulic fracturing fluids and provide the product name to the state, rather than have to submit separate analyses. Alabama regulations apply this approach where fracturing would occur within an underground source of drinking water.

EPA's 2004 Review of Hydraulic Fracturing for CBM Production

In response to the *LEAF I* decision, citizen reports of water well contamination attributed to hydraulic fracturing of coal beds, and the rapid growth in CBM development, EPA undertook a study to evaluate the environmental risks to underground sources of drinking water from hydraulic fracturing practices associated with CBM production. EPA issued a draft report in August 2002.[70] The draft report identified water quality and quantity problems that individuals had attributed to hydraulic fracturing of coal beds in Alabama, New Mexico, Colorado, Wyoming, Montana, Virginia, and West Virginia.[71] Based on the preliminary results of the study, EPA tentatively concluded that the potential threats to public health posed by hydraulic fracturing of coalbed methane wells appeared to be small and did not justify additional study or regulation.

EPA also reviewed whether direct injection of fracturing fluids into underground sources of drinking water posed any threat. EPA reviewed 11 major coalbed methane formations to determine whether coal seams lay within

USDWs. EPA determined that 10 of the 11 producing coal basins "definitely or likely lie entirely or partially within USDWs."

In January 2003, the EPA's National Drinking Water Advisory Council submitted to the EPA Administrator a report on hydraulic fracturing, underground injection control, and coalbed methane production and its impacts on water quality and water resources. The Council noted concerns regarding: (1) the lack of resources to implement the UIC program, (2) the use of diesel fuel and potentially toxic additives in the hydraulic fracturing process, (3) the potential impact of coalbed methane development on local underground water resources and the quality of surface waters, and (4) the maintenance of EPA regulatory authority within the UIC program. The Council recommended that EPA:

- work through regulatory or voluntary means to eliminate the use of diesel fuel and related additives in fracturing fluids that are injected into formations containing sources of drinking water;
- continue to study the extent and nature of public health and environmental problems that could occur as a result of hydraulic fracturing for coalbed methane production; and
- defend EPAs existing authority and discretion to implement the UIC program in a manner that advances the protection of ground water resources from contamination.[72]

In 2004, EPA issued a final version of the 2002 draft report, based primarily on an assessment of the available literature and extensive interviews, and concluded that the injection of hydraulic fracturing fluids into CBM wells posed little threat to underground sources of drinking water and required no further study. However, EPA found that very little documented research had been done on the environmental impacts of injecting fracturing fluids.[73] Additionally, EPA had discussed the use of diesel fuel in fracturing fluids in the 2002 draft report, and concluded in the final report that:

> The use of diesel fuel in fracturing fluids poses the greatest potential threat to USDWs because the BTEX constituents in diesel fuel exceed the MCL [maximum contaminant level] at the point-of-injection.[74]

EPA also noted that estimating the concentration of diesel fuel components and other fracturing fluids beyond the point of injection was beyond the scope of its study.[75] Moreover, the EPA study focused specifically on CBM wells and did not review the use of hydraulic fracturing in other geologic formations, such as the Marcellus and Barnett shales and tight gas sand formations.

To address concerns about the use of diesel fuel in hydraulic fracturing fluids, EPA entered into an agreement with three companies that provided roughly 95% of hydraulic fracturing services (BJ Services, Halliburton Energy Services, and Schlumberger Technology Corporation). Under this agreement, the firms agreed to remove diesel fuel from CBM fluids injected directly into drinking water sources if cost-effective alternatives were available.[76]

EPACT 2005: A LEGISLATIVE EXEMPTION FOR HYDRAULIC FRACTURING

The decision by the U.S. Court of Appeals for the Eleventh Circuit in *LEAF I* highlighted a debate over whether the SDWA as it read at the time required EPA to regulate hydraulic fracturing. Although the Eleventh Circuit's decision applied only to hydraulic fracturing for coalbed methane production in Alabama, the court's reasoning—in particular, its finding that hydraulic fracturing "unquestionably falls within the plain meaning of the definition [of underground injection]"[77]—raised the issue of whether EPA could be required to regulate hydraulic fracturing under the SDWA.

Before this question was resolved through agency action or litigation, Congress passed an amendment to the SDWA as a part of EPAct 2005 (P.L. 109-58) that addressed this issue. Section 322 of EPAct 2005 amended the definition of "underground injection" in the SDWA as follows:

> The term "underground injection"—(A) means the subsurface emplacement of fluids by well injection; and (B) excludes—(i) the underground injection of natural gas for purposes of storage; and (ii) the underground injection of fluids or propping agents (other than diesel fuels) pursuant to hydraulic fracturing operations related to oil, gas, or geothermal production activities.

This amendment clarified that the UIC requirements found in the SDWA do not apply to hydraulic fracturing, although the exclusion does not extend to the use of diesel fuel in hydraulic fracturing operations. This amended language is the definition of "underground injection" found in the SDWA as of the date of this report.

Roughly five years after enactment of this amendment, EPA posted on its website a requirement that any service company that performs hydraulic fracturing using diesel fuel must receive prior authorization from the relevant UIC authority (state or EPA). EPA also determined that injection wells receiving diesel fuel are Class II wells for purposes of the UIC program. These determinations were not made via adoption of new regulations, but rather were noted on the agency's website in 2010.[78] This determination on the EPA website (without notice and comment as would normally be done in an agency rulemaking proceeding) has been challenged by oil and gas trade organizations in a petition for review in the U.S. Court of Appeals for the District of Columbia.[79]

Proposed Legislation in the 112th Congress: The FRAC Act

On March 15, 2011, the Fracturing Responsibility and Awareness of Chemicals Act of 2011 (FRAC Act), H.R. 1084 and S. 587, was introduced in both the Senate and the House of Representatives.[80] The bills have some minor language differences, but are substantially similar. (They also are similar to bills introduced in the past Congress.) Each contains two amendments to the SDWA—one that would amend the definition of underground injection to include hydraulic fracturing, and another that would create a new disclosure requirement for the chemicals used in hydraulic fracturing.

H.R. 1084 provides that the definition of "underground injection" that was amended in 2005 to exclude most hydraulic fracturing would be amended once again to include "the underground injection of fluids or propping agents pursuant to hydraulic fracturing operations related to oil, gas or geothermal production activities," excluding injection of natural gas for subsurface storage.[81] This would not only repeal the amended definition of "underground injection" that was enacted as part of EPAct 2005 which excluded hydraulic fracturing, but would essentially codify the court's decision in *LEAF I* and clear up any ambiguity regarding regulation of hydraulic fracturing under the SDWA.

The second amendment to the SDWA in the FRAC Act would create a new hydraulic fracturing disclosure requirement. H.R. 1084 would create a new statutory obligation requiring anyone conducting hydraulic fracturing to:

Disclose to the State (or the [EPA] if the [EPA] has primary enforcement responsibility in the State)—(I) prior to the commencement of any hydraulic fracturing operations at any lease area of portion thereof, a list of chemicals intended for use in any underground injection during such operations, including identification of the chemical constituents of mixtures, Chemical Abstracts Service numbers for each chemical and constituent, material safety data sheets when available, and the anticipated volume of each chemical; and (II) not later than 30 days after the end of any hydraulic fracturing operations the list of chemicals used in each underground injection during such operations, including identification of the chemical constituents of mixtures, Chemical Abstracts Service numbers for each chemical and constituent, material safety data sheets when available, and the volume of each chemical used.[82]

The bill would also require that the state or EPA "make the disclosure of chemical constituents ... available to the public, including by posting the information on an appropriate Internet Web site," and the bill clarifies that the disclosure requirements "do not authorize the State (or the [EPA]) to require the public disclosure of proprietary information."[83] In other words, the disclosure requirements address only the chemicals used, not the manner of their use or the amounts or ratios in which they are used. This language attempts to protect proprietary business information, that is, "secret" formulas or practices that drilling companies may feel they should not be required to disclose to their competitors. Some state oil and gas production statutes and regulations extend similar protections for proprietary business information, while still requiring disclosure to regulators of the chemical constituents being used in hydraulic fracturing.[84]

Furthermore, the FRAC Act would require operators to disclose proprietary chemical information to treating medical professionals in cases of medical emergencies.[85] Although most state oil and gas rules do not require disclosure of proprietary chemical information to medical professionals, such disclosure broadly parallels federal requirements under the Occupational Safety and Health Act (OSHAct).[86] Nonetheless, the OSHAct requirements were not designed for environmental investigation purposes and have been criticized as deficient. Calls for disclosure of hydraulic fracturing chemicals have increased as homeowners and others express concern about the potential presence of unknown chemicals in tainted well water near oil and gas operations.

Potential Implications of Hydraulic Fracturing Regulation under the SDWA

The full regulation of hydraulic fracturing under the SDWA (i.e., beyond injections involving diesel) potentially could have significant, but currently unknowable, environmental benefits as well as impacts on oil and natural gas producers and state and federal regulators. Resulting groundwater protection, public health, and economic benefits would likely be experienced most significantly in states that may currently have relatively weak groundwater protection requirements (such as substandard cementing and casing requirements, or permitting injection of unknown chemicals directly into USDWs). However, the specific regulation of the underground injection of fluids for hydraulic fracturing purposes would not address surface management of chemicals, drilling wastes, or treatment and disposal of produced water. If such surface activities are determined to be the sources of most water contamination incidents associated with unconventional oil and gas development, federal regulation of hydraulic fracturing under the SDWA may have limited environmental and public health benefits. Benefits of federal regulation could be significant if various states have oil and gas regulations that are not adequate to address the new drilling and production methods applied to unconventional oil and gas resources.

Regulations requiring chemical disclosure could also be beneficial. The lack of information regarding chemicals used in hydraulic fracturing has made investigations of contamination difficult as well owners and state regulators typically do not know which chemicals to test for to determine whether a fracturing fluid has migrated into a water source.[87]

If the SDWA were amended to authorize (but not mandate) EPA to regulate hydraulic fracturing, EPA likely would need to undertake further study to assess the potential risks of hydraulic fracturing to underground sources of drinking water. (The Agency currently is conducting such studies, as discussed below.) Subsequently, EPA might determine the need for, and potential scope of, any new regulations, and decide whether to adapt the existing regulatory framework or to develop a new approach under the UIC program. The rulemaking process typically takes several years. A 2009 presentation by EPA's Region 8 UIC program explained that, if legislative change occurs:

Additional study may take place, regulations may be written by EPA, some combination of these may happen, [and] there may be a phased-in approach. If regulations are developed, they typically include: establishing a regulation development workgroup which can include the public; a proposed regulation, including opportunity for public comment (and one or more hearings if needed); a final regulation, including opportunity for judicial appeals; and an effective date for the regulation.[88]

One implication of regulating hydraulic fracturing under SDWA relates to the SDWA's citizen suit provisions. As noted, Section 1449 provides for citizen civil actions against any person or agency allegedly in violation of provisions of SDWA, or against the EPA Administrator for alleged failure to perform any action or duty that is not discretionary.[89] This provision could represent an expansion in the ability of citizens to challenge state administration of oil and gas programs related to hydraulic fracturing and drinking water, were the hydraulic fracturing exemption provision to be repealed.

As discussed, the SDWA currently includes two options for approving state UIC programs related to oil and gas recovery.[90] Under the less restrictive requirements of Section 1425, EPA may be able to implement new requirements primarily through guidance and review and approval of state programs revised to address hydraulic fracturing. EPA used this approach when ordered to require Alabama to regulate hydraulic fracturing of coal beds, and a federal district court approved this approach.

If EPA decided to allow states to regulate hydraulic fracturing under Section 1425, the agency also might be expected to write new hydraulic fracturing regulations under Section 1421 in order to implement the program directly for states that do not have primacy for Class II wells and for states that exercise primacy under Section 1422. Regardless of regulatory approach, new requirements would likely require substantially more resources for UIC program administration and enforcement by the states and EPA.

The possible impacts of enacting legislation directing EPA to regulate hydraulic fracturing could vary for different oil and gas production operations. The SDWA directs EPA, when developing UIC regulations, to take into consideration "varying geologic, hydrological, or historical conditions in different States and in different areas within a State."[91] Consequently, if EPA were to regulate hydraulic fracturing under the SDWA, the agency conceivably could establish different requirements to address such differences among states or regions. If practical and applicable, EPA might find this statutory flexibility helpful, as the USDW contamination risks of hydraulic fracturing could vary widely among different formations and settings. For example, fracturing a coal bed that may qualify as a USDW poses very different groundwater contamination risks than fracturing a shale formation that is widely separated from any USDW.[92] Thus, the possible application and impact of federal regulations might vary significantly in different formations, and the impacts and potential environmental benefits would likely be greatest in such coal beds or other formations occurring in USDWs. Arguably, certain groundwater protection requirements might not be needed for some shale formations, such as much of the Marcellus Shale, because these formations generally lie deep beneath most USDWs.[93] However, EPA might broadly apply other requirements, such as those related to well drilling and construction, to protect USDWs through which wells may pass, among other purposes.

For the oil and gas industry, regulation of hydraulic fracturing under the UIC program could have a range of impacts. In some states, oil and gas operations are subject to regulation by a state oil and gas agency or commission as well as an environmental or public health agency. Industry representatives have expressed concern over potential for some duplication of requirements from state oil and gas regulators and environmental regulators. Delays in issuing permits, and commensurate delays in well stimulation and gas marketing are among the concerns. The citizen suit provision of the SDWA also may be an issue. One analysis attempting to measure the economic and energy effects of potential regulation noted that:

> Experience suggests that there will be a reduction in the number of wells completed each year due to increased regulation and its impact on the additional time needed to file permits, push-back of drilling schedules due to higher costs, increased chance of litigation, injunction or other delay tactics used by opposing groups and availability of fracturing monitoring services.[94]

Several studies have attempted to estimate the potential economic and energy supply impact of regulating hydraulic fracturing under the federal UIC program. A 2009 study prepared by a consultant for the DOE estimated the costs associated with "a stringent set of potential federal requirements" including (1) obtaining a permit, (2) conducting an area of review assessment, (3) performing in-situ stress analysis, (4) conducting three-dimension fracture simulation, (5) monitoring, (6) mapping fractures, or conducting other post-fracture analysis, (7) for some wells (perhaps 10%), performing state-of-the-art down-hole fracture imaging, and (8) additional cement to ensure isolation of the target zone before fracturing.[95] Based on these assumed elements of a regulatory program, the study estimated that the compliance costs for regulating hydraulic fracturing for oil and gas development would be $100,505 for new wells receiving hydraulic fracturing treatment.[96]

A stringent regulatory program under Section 1422 arguably could include many of the above requirements. However, it is unknown what EPA might require and unclear what costs would be attributed to federal regulation. Some activities already are used in the industry or required by states (e.g., well cementing across all groundwater zones). EPA UIC staff note that some of the requirements assumed in the study have never been a part of the federal UIC regulations. Other effects that are not easily quantified include the costs associated with waiting periods between fracturing jobs for approvals and other potential disruptions to operations.

The Ground Water Protection Council (GWPC),[97] representing state agencies, has opposed reclassification of hydraulic fracturing as a permitted activity under the UIC programs, stating that (1) a risk has not been identified, and thus, there is no evidence that regulation is necessary; and (2) UIC regulation would divert resources from higher risk activities.[98] The legislatures of major oil and gas producing states, including the states of Alabama, Alaska, Montana, North Dakota, Wyoming, and Texas, passed and sent to Congress resolutions asking Congress not to extend SDWA jurisdiction over hydraulic fracturing activities.

As discussed, the GWPC is recommending the adoption of various best management practices to strengthen protection of water resources when developing oil and gas resources. Industry appears to be adopting some of these management practices independent of regulation. In discussing lessons learned from developing the Barnett shale, industry consultants recently reported, that an "important factor, requiring 3D seismic [imaging], is the avoidance of geo-hazards, such as water-bearing karsts and faults."[99] However, voluntary industry practices can not be enforced, and there is no assurance that they would be widely adopted.

If EPA were to regulate hydraulic fracturing under the UIC program, many public concerns regarding the development of shale gas or coalbed methane might persist and be of greater concern than the potential for contamination of USDWs through the fracturing process. These concerns involve land surface disturbances associated with the development of roads, well pads, and natural gas gathering pipelines; potential impacts of water withdrawal; treatment and disposal of flowback water to surface waters; air quality impacts; etc. Some of these activities are subject to other federal laws, such as Clean Water Act requirements covering the treatment and discharge of produced water into surface waters. The regulatory impacts of state and federal regulatory requirements for treatment and discharge of produced water may be more significant than potential UIC-related requirements. Other impacts related to development of unconventional oil and gas resources are highly visible and may cause as much or more potential problems and public concern than the specific process of deep underground fracturing of oil and gas formations and coal beds. Some of these issues (particularly certain land-use and mineral resource development issues) are beyond the reach of federal regulation, and thus, are left to state and local governments to address.[100] New York State's draft Supplemental Generic Environmental Impact Statement is one example of a state taking a comprehensive approach to addressing a broad range of possible environmental impacts that could be associated with Marcellus Shale development.[101]

STATE REGULATION OF HYDRAULIC FRACTURING

While the federal government currently exempts most hydraulic fracturing activity from regulation under the SDWA, the states are free to regulate the practice as they see fit. Although state oil and gas regulatory programs initially focused on managing petroleum reservoirs, efficient production, and addressing mineral rights issues, these programs have become more environmentally focused through the decades. The GWPC and the Interstate Oil and Gas Compact Commission (IOGCC)[102] each report that the major oil and gas producing states now have laws and regulatory requirements in place to protect water resources during oil and natural gas exploration and production activities.

Both the GWPC and the IOGCC oppose federal regulation of hydraulic fracturing, noting that this process is regulated by the states, sometimes specifically, but most often through general oil and gas production regulations, policies, and practices.[103] The IOGCC notes that member states have adopted comprehensive laws and regulations to provide for safe operations and to protect the nation's drinking water sources, and that these states have trained personnel with expertise to effectively regulate oil and gas exploration and production; thus, making the states the best-suited regulators of hydraulic fracturing. The IOGCC further makes the case for keeping responsibility with the states:

Hydraulic fracturing is currently, and has been for decades, a common operation used in exploration and production by the oil and gas industry in all gas producing states. Because of the unique position of the states and their collective expertise on matters concerning the oil and gas industry, regulation of hydraulic fracturing should remain the responsibility of the States. The States have as much of a vested interest in the protection of groundwater as the federal government and as such, will continue to regulate the process effectively and efficiently, taking into account the particulars of the geology and hydrology within their boundaries. There is not a "one-size fits all" approach to effective regulation.[104]

The question that has arisen is whether state oil and gas programs effectively address increasing groundwater protection concerns arising with the heightened concentration and broadened geographic extent of unconventional and conventional oil and gas resource development that relies on hydraulic fracturing in combination with deep horizontal drilling.[105] Several states, including Pennsylvania and New York, currently are reviewing and/or revising their oil and gas exploration and production regulations in response to new types and levels of natural gas production. Colorado and Wyoming recently amended their oil and gas rules to increase protection of water resources.

A related issue concerns the extent to which state oil and gas agencies coordinate adequately with their water pollution control counterparts. Most states have different agencies administering oil and gas programs and environmental programs. State UIC programs often are administered by the environmental agency, while oil and gas exploration and production activities are overseen by separate oil and gas entities. Moreover, with the exception of Alabama, which acted in response to a court ruling, no state chose to regulate hydraulic fracturing as part of its EPA-authorized underground injection control program prior to the 2005 amendment to the SDWA.[106]

GWPC Review of State Regulations

Although states have extensive regimes in place to manage oil and gas development activities, the GWPC also notes that related state groundwater protection regulations, policies, and practices can be uneven. The GWPC recently reviewed, for the 27 major oil and gas producing states, state oil and gas regulations designed to protect water resources.[107] Based on this review, the GWPC concluded that, in general, state oil and gas regulations are adequately designed to protect water resources. Among the states, regulatory requirements to protect water resources address permitting, well drilling and construction (e.g., casing, cementing, and test pressure requirements), well closure and abandonment, and waste fluid management.

While few states explicitly mention hydraulic fracturing in their regulations, many have well drilling, construction, completion, and reporting requirements intended to protect ground and surface water resources. For example, 10 major producing states require reporting of chemicals used in well treatments, 25 states require operators to submit well treatment (including fracturing) reports, and 22 states require operators to cement across groundwater zones. State requirements vary greatly, from the detailed requirements in Alabama, to more general mandates not to harm water resources (e.g., Arizona oil and gas rules require operators to "conduct operations in a manner that prevents oil, gas, salt water, fracturing fluid or any other substance from polluting any surface or subsurface waters"). Colorado's regulations include a well casing program to protect groundwater (and hydrocarbons), require well treatment and fracturing reporting, and require operators to notify landowners at least one week before conducting various operations, including fracturing.[108]

While finding that most states have an extensive array of permitting and operating requirements for oil and gas wells, the GWPC also noted that some states lacked important provisions in their programs. For example, most, but not all, states have well construction requirements that include provisions for cementing above oil or gas producing zones and across groundwater zones. The GWPC made a series of recommendations to strengthen state programs to protect water resources. A sample of findings and recommendations from the GWPC review follows:

- State oil and gas regulations are adequately designed to directly protect water resources through the application of specific program elements such as permitting, well construction, well plugging, and temporary abandonment requirements.
- Some exploration and production (E&P) activities have caused contamination of both surface and ground water. Past practices related to pit construction, well cementing and operation, and well plugging were not always adequate to prevent migration of contaminants to surface and ground water.

- States should review current regulations in several program areas to determine whether they meet an appropriate level of specificity (e.g., use of standard cements, plugging materials, pit liners, siting criteria, and tank construction standards, etc.).
- Comprehensive studies should be undertaken to determine the relative risk to ground water resources from the practice of shallow hydraulic fracturing. These studies should be used, with current knowledge, to develop a generic set of best management practices (BMPs) for hydraulic fracturing which state agencies could use either to develop state specific BMPs or develop additional state regulations.
- Hydraulic fracturing in oil or gas bearing zones that occur in non-exempt USDWs should be either stopped, or restricted to the use of materials that do not pose a risk of endangering ground water and do not have the potential to cause human health effects (e.g., fresh water, sand, etc.).
- Hydraulic fracturing of deep zones poses little to no risk of groundwater contamination.
- Many states split jurisdiction between oil and gas and water quality or pollution control agencies over some aspects of oil and gas regulation including tanks, pits, waste handling and spills. Where split jurisdiction of oil and gas operations exists, formal memoranda of agreement and regulatory implementation plans should be negotiated.[109]
- States should consider requiring companies to submit a list of additives used in formation fracturing and their concentration within the fracture fluid matrix. Further, states that do not currently regulate handling and disposal of fracture fluid additives and constituents recovered during recycling operations should consider the need to develop such regulations.
- A state program review process, conducted by the national nonprofit group, State Review of Oil and Natural Gas Environmental Regulations (STRONGER),[110] should be recognized as an effective tool for assessing the capability of state programs to manage E&P waste and measure program improvement over time.
- Best Management Practices that can be adapted to each state should be developed to manage hydraulic fracturing. STRONGER should evaluate whether to update its mission to include environmental elements of state oil and gas programs beyond the traditional area of E&P waste [to include hydraulic fracturing]. (STRONGER issued

hydraulic fracturing guidelines in February 2010, and review teams are using the guidelines to the evaluate oil and gas regulatory programs of states that have volunteered to be reviewed.

Various states have determined that the expanding development of unconventional oil and gas resources, along with the increased use of hydraulic fracturing and directional drilling, requires more state oversight. Some states are responding by increasing staff resource levels. And in some states, such Colorado, New York and Pennsylvania, this expansion has prompted a reassessment of the adequacy of current regulations and policies.[112]

UIC Program Resources

The funding and staffing resource implications of including hydraulic fracturing under the UIC program could be significant for regulatory agencies. The scope of the added workload under Class II UIC programs could more than double. Currently, there are approximately 146,800 Class II wells nationwide.[113] In contrast, the DOE Energy Information Administration reports that the number of gas producing wells in the United States increased from 302,421 in 1999 to 493,100 wells in 2009, and most new wells are fractured.[114]

EPA's annual appropriation includes funds for state grants to support state administration of various EPA programs. Since the 1980s, annual appropriations to support state UIC programs have remained essentially flat (not accounting for inflation) at roughly $10.5 million.[115] Ten EPA regional offices and 42 states share this amount annually to administer the entire UIC program, which covers 1.7 million wells (Classes I through V) nationwide.[116] The GWPC has estimated that annual UIC program funding would need to increase to $56 million to fully meet the needs of the existing UIC program.[117] The GWPC further estimated that EPA would need to provide funding at a level of $100 million annually to meet the needs for the full UIC program, including the regulation of geologic sequestration of carbon dioxide. Given the large number of wells that are fractured, UIC program oversight and enforcement costs for state agencies could be considerably higher if this process is subjected to federal UIC regulations. EPA and states would need to develop new regulatory requirements for these wells and increase staff to review applications and make permitting decisions. States and industry representatives have expressed concern that failure to provide sufficient resources would likely create permitting backlogs. For example, under UIC regulations, EPA or the primacy state must provide for a public hearing for each permit issuance, and have inspectors on site.[118] Some states impose permit fees or use other revenue-generating mechanisms, while such approaches have not been embraced in other states.

Because of the sheer number of potentially newly regulated wells, EPA (given its current resource levels) would necessarily need to rely heavily upon the states to implement this program. In 2007, the GWPC noted that states are already struggling to fully implement their UIC programs, and new requirements for hydraulic fracturing, in addition to EPA's pending UIC regulation for geologic sequestration of carbon dioxide, would be problematic. Specifically, the GWPC cautioned that without substantial increases in funding for the UIC program:

- More states will decide to return primacy to EPA (which also would not have additional funds to implement the program).
- The overall effectiveness of UIC programs will suffer as more wells and well types are added without a concurrent addition of resources to manage them.
- Decisions regarding which parts of the program to fund with limited dollars could result in actual damage to USDWs if higher risk/higher cost portions of the program are put "on the back burner."
- Negative impacts on the economy could occur as permitting times lengthen due to increased program workloads.[119]

EPA resources are also at issue. The agency would require additional technically trained staff to oversee and enforce state programs and implement the program in non-primacy states. Should some states decide not to assume primacy for the new program, EPA's resource challenges would increase. As with states, EPA resources are stretched. For example, the agency is continuing its review and approval of various state Class V UIC programs that are being revised to implement a 1999 rulemaking. Additionally, EPA proposed a rule in 2008 to establish a new Class VI well for the geologic sequestration of carbon dioxide. The agency is working to finalize this UIC rule.

STUDIES AND RESEARCH

Technical and practical questions regarding the development of the unconventional oil and gas resources, such as the Marcellus Shale, remain unanswered. USGS researchers have noted that while drilling and hydraulic fracturing technologies have improved over the past several decades, "the knowledge of how this extraction might affect water resources has not kept pace."[120] Consequently, environmental regulators and gas developers face new challenges and some uncertainties as the Marcellus Shale is developed.

Several studies and research projects are underway related to hydraulic fracturing for the purpose of oil and gas development. In August 2009, the DOE announced that it was funding nine new research projects intended to improve methods for treating, reusing, and managing water associated with natural gas development - including gas from coal beds and shale. Several of these projects address hydraulic fracturing, including projects to develop processes and technologies for pretreatment of produced brine and hydraulic fracturing flowback waters.

Another project is intended to develop a new hydraulic fracturing module to assist regulators and operators in enhancing protective measures for source water and streamlining the well-permitting process.[121] Such research studies could help reduce water contamination risks associated with fracturing and reduce regulatory impacts.

In EPA's FY2010 appropriations act (P.L. 111-88), Congress directed EPA to carry out a study on the relationship between hydraulic fracturing and drinking water, using a credible approach that relies on the best available science, as well as independent sources of information.[122] EPA expects to report on the interim research results in 2012, and issue a follow-up report in 2014. EPA's Draft Hydraulic Fracturing Study Plan states that the overall purpose of the study is to understand the relationship between hydraulic fracturing and drinking water resources. Specifically, the study is designed to examine conditions that may be associated with potential contamination of drinking water sources, and to identify factors that may lead to human exposure and risks. EPA has proposed research studies that address the full lifecycle of water in hydraulic fracturing, from water acquisition and chemical mixing, through actual fracturing and to post-fracturing stages, including the management of flowback and produced water and its treatment and/or disposal.[123]

As part of the study, EPA plans to investigate existing reported incidents of drinking water resource contamination where hydraulic fracturing has occurred. These retrospective case studies will be used to determine the potential relationship between reported impacts and hydraulic fracturing activities. Also, prospective case studies will include sampling and water resource characterization before fracturing occurs, and then evaluate any water quality or chemistry changes subsequently. The EPA studies may add insight regarding the risks of hydraulic fracturing as well as specific regulatory gaps and needs.

CONCLUDING OBSERVATIONS

Hydraulic fracturing bills introduced in the 112th Congress and previously have generated considerable debate. Industry and many state agencies are arguing against regulation of hydraulic fracturing under the SDWA, and note a long history of the successful use of this practice in developing oil and gas resources. Industry representatives argue that additional federal regulation is unnecessary and would likely slow domestic gas development and increase energy prices. At the same time, the amount of natural gas produced from unconventional and conventional formations that rely on hydraulic fracturing continues to grow, and now provides the dominant share of domestically produced natural gas. However, drilling and fracturing methods and technologies have changed significantly over time as they are applied to more challenging formations, increasing markedly the amount of water and fracturing fluids involved in production operations. It is the rapidly increasing and geographically expanding use of hydraulic fracturing, along with a growing number of citizen complaints of groundwater contamination and other environmental problems attributed to this practice, that has led to calls for greater state and/or federal environmental oversight of this activity.

The central issue in the debate has concerned the need for, and potential benefits of, federal regulation of hydraulic fracturing. Pollution prevention generally, and groundwater protection in particular, is much less costly than cleanup, and where groundwater supplies are not readily replaceable, protection becomes a high priority. Environmental regulations generally involve internalizing costs associated with processes. And federal regulations generally are used to address activities found to have widespread public health and environmental risks, particularly where significant regulatory gaps and unevenness exists among the states. To the extent that a regulation is needed and is well designed and implemented, public benefits (i.e., protecting water resources) would be expected to accrue. If EPA were to regulate fracturing, the environmental benefits could be significant if the risks of contamination are significant. Alternatively, the benefits may be small if most pollution incidents are found to be related to other oil and gas production activities, such as improper disposal of produced water or mishandling of materials on the surface. Some of these issues are not subject to SDWA authority and would not be addressed through regulation under this act.

State oil and gas and groundwater protection agencies widely support keeping responsibility for regulating hydraulic fracturing with the states. In September 2009, the GWPC approved a resolution supporting continued state regulation of hydraulic fracturing and encouraging Congress, EPA, DOE, and others to work with the states and the GWPC to evaluate the risks posed by hydraulic fracturing. The GWPC and others have expressed concern that regulation of hydraulic fracturing under the SDWA would divert compliance and enforcement resources from higher priority issues. Additionally, the IOGCC has adopted a resolution urging Congress not to remove the fracturing exemption from provisions of the SDWA, noting that the process is a temporary injection-and-recovery technique and does not fit the UIC program which EPA generally developed to address the permanent disposal of wastes.

Nonetheless, given the importance of good quality water supplies to homeowners, ranchers, and communities, and uneven regulation across the states, some continue to urge a federal solution. It could be expected that the potential impact of federal regulations on states and industry would be mitigated (or provide fewer added benefits) to the degree that states currently have pertinent and effective requirements, or respond to increased development of unconventional gas and oil resources with their own increased requirements. A range of new regulatory requirements is under consideration or in place in various states.

Whether state or federal, regulations require adequate implementation resources to be administered effectively. The regulation of hydraulic fracturing under the SDWA could pose significant new staffing and other resource demands on EPA and the states, although those states that have effective requirements in place to address hydraulic fracturing may not experience significant impacts.

Currently, there is little agreement as to the risks to underground sources of drinking water associated with hydraulic fracturing. Given the statements of EPA and state groundwater protection representatives, it appears that the actual and potential risks associated with hydraulic fracturing could benefit from a better scientific understanding.

Congress has directed EPA to study the relationship between hydraulic fracturing and drinking water. The results of the pending study may help inform the need for additional regulation, whether at the state or federal level. The EPA study and other federal research, along with work being done in the states, should enable a better assessment of the risks that fracturing may pose to drinking water resources, and could help inform any potential congressional action as well as the agency's subsequent response if legislation were enacted.

End Notes

[1] Hydraulic fracturing is also used for other purposes, such as developing water supply wells and geothermal production wells. This report focuses only on its use for oil and gas development.

[2] Tight gas sands are sandstone formations with very low permeability that must fractured to release the gas.

[3] According to the Schlumberger *Oilfield Glossary*, propping agents, or proppants, are "sized particles mixed with fracturing fluid to hold fractures open after a hydraulic fracturing treatment. In addition to naturally occurring sand grains, man-made or specially engineered proppants, such as resin-coated sand or high-strength ceramic materials like sintered bauxite, may also be used." The glossary is available at http://www.glossary.oilfield.slb.com/default.cfm.

[4] This process is distinct from enhanced oil and gas recovery and other secondary and tertiary hydrocarbon recovery techniques which involve separate wells. Injections for hydraulic fracturing are done through the production wells.

[5] The Schlumberger glossary notes that "produced fluid is a generic term used in a number of contexts but most commonly to describe any fluid produced from a wellbore that is not a treatment fluid. The characteristics and phase composition of a produced fluid vary and use of the term often implies an inexact or unknown composition." "Flowback" refers to "the process of allowing fluids to flow from the well following a treatment, either in preparation for a subsequent phase of treatment or in preparation for cleanup and returning the well to production."

[6] U.S. Department of Energy, Office of Fossil Energy and National Technology Laboratory, *Modern Shale Gas Development in the United States: A Primer*, DE-FG26-04NT15455, April 2009, p. 66, http://fossil.energy.gov/programs/oilgas/publications/ naturalgas_general/ Shale_Gas_Primer_2009.pdf.

[7] American Petroleum Institute, Hydraulic Fracturing, http://www.api.org/policy

[8] U.S. Environmental Protection Agency, Study Design for Evaluation of Impacts to Underground Sources of Drinking Water by Hydraulic Fracturing of Coalbed Methane Reservoirs, http://www.epa.gov/safewater/uic/ wells_coalbedmethanestudy_ finalstudy design.html.

[9] U.S. Environmental Protection Agency, *Effluent Guidelines: Coalbed Methane Extraction Detailed Study*, http://water

[10] DOE reports that proved reserves of shale gas increased from 21,735 billion cubic feet (bcf) in 2007 to 32,825 bcf in 2008. U.S. Department of Energy, Energy Information Agency, *Natural Gas Navigator: Shale Gas Proved Reserves*, October 29.2009, http://tonto.eia.doe.gov/dnav/ng/ng_enr_shalegas_s1_a.htm.

[11] U.S. Energy Information Administration, *Natural Gas Navigator: Number of Producing Gas Wells*, August 2009, http://tonto.eia.doe.gov/dnav/ng/ng_prod_wells

[12] Ground Water Protection Council and ALL Consulting, *Modern Shale Gas Development in the United States: A Primer*, U.S. Department of Energy, Office of Fossil Energy and National Energy Technology Laboratory, April 2009, pp. 47-48, http://www.netl.doe.gov/technologies/oil-gas/publications/EPreports/Shale_Gas_Primer_2009.pdf.

[13] Multiple fractures are typical in deep shale formations. Scott Stevens and Vello Kuuskraa of Advanced Resources International report that "[t]oday, deep shale drillers all employ essentially the same Barnett-style well drilling and completion design: ±4,000-ft. long lateral stimulated by multimillion-lb slick-water fracs in a dozen stages." Source: Seven Plays Dominate North America Activity, *Oil and Gas Journal*, September 28, 2009, v. 107, n. 36 p. 41.

[14] Using slickwater fracturing increases the rate at which fluid can be pumped down the wellbore to fracture the shale. The process may involve the use of friction reducers, biocides, surfactants, and scale inhibitors. Biocides prevent bacteria from clogging wells; surfactants help keep the sand or other proppant suspended. Slickwater fracturing was first used in the Barnett shale in Texas.

[15] The scope of this report is limited to potential issues related to hydraulic fracturing and contamination of underground sources of drinking water related to the fracturing process. Another environmental concern related to hydraulic fracturing is the disposal or treatment of "flowback" from the fracturing/drilling process, which may present environmental and regulatory issues and also water treatment infrastructure issues. Disposal of flowback by means other than disposal through injection wells would be regulated pursuant to the Clean Water Act. For a discussion of the hydraulic fracturing process and potential sources of water contamination, including surface water contamination, see CRS Report R40894, *Unconventional Gas Shales: Development, Technology, and Policy Issues*, coordinated by Anthony Andrews. For a discussion of the "discharge" requirements under the Clean Water Act, see EPA, *Natural Gas Drilling in the Marcellus Shale: NPDES Program Frequently Asked Questions*, March 16, 2011, http://www.epa.gov/npdes/pubs/hydrofracturing_faq.pdf.

[16] The GWPC is a national association representing state groundwater and UIC agencies whose mission is to promote protection and conservation of groundwater resources for beneficial uses. The stated purpose of the GWPC is "to promote and ensure the use of best management practices and fair but effective laws regarding comprehensive ground water protection." http://www.gwpc.org

[17] Ground Water Protection Council, U.S. Department of Energy, Office of Fossil Energy, National Energy Technology Laboratory, *State Oil and Natural Gas Regulations Designed to Protect Water Resources*, May 2009, p. 24. Coal beds are often a source of good quality groundwater, thus, presenting challenges to developers and potential conflicts with well owners.

[18] For a discussion of environmental concerns and recommendations, see, for example, Environmental Working Group, *Drilling Around the Law*, January 2010, http://static.ewg.org/files/EWG-2009drillingaroundthelaw.pdf.

[19] Pennsylvania Department of Environmental Protection, Consent Order and Agreement, November 4, 2009; http://s3.amazonaws.com/propublica/assets

[20] CERCLA, § 104(b), authorizes the President to undertake investigations, monitoring, surveys, testing, or other investigations to identify the existence and extent of a release, the source and nature of pollutants involved, and the extent of danger to public health, welfare, or the environment, "whenever the President has reason to believe a release has occurred, or is about to occur, or that illness, disease, or complaints thereof may be attributable to exposure to a hazardous substance, pollutant, or contaminant." (42 U.S.C. § 9604(b)).

[21] U.S. Environmental Protection Agency, Region 8, *Groundwater Investigation: Pavillion*, http://www.epa.gov/region8/superfund/wy/pavillion/index.html

[22] P.L. 111-88, H.Rept. 111-316.

[23] The Safe Drinking Water Act of 1974 (P.L. 93-523) authorized the UIC program at EPA. UIC provisions are contained in SDWA Part C, §§ 1421 - 1426; 42 U.S.C. §§ 300h - 300h-5.

[24] 42 U.S.C. § 300h(b)(2).

[25] P.L. 109-58, § 322.

[26] 42 U.S.C. § 300h(d).

[27] *Id.*

[28] 42 U.S.C. § 300h-1. The minimum requirements for a state UIC program can be found at 40 C.F.R. Part 145.

[29] 42 U.S.C. § 300h-4. SDWA § 1425 was added by the Safe Drinking Water Act Amendments of 1980, P.L. 96-502. The House committee report accompanying the legislation that added § 1425 noted that:
Most of the 32 states that regulate underground injection related to the recovery or production of oil or natural gas (or both) believe they have programs already in place that meet the minimum requirements of the Act including the prevention of underground injection which endangers drinking water sources. This is especially true of the major producing states where underground injection control programs have been underway for years. It is the Committee's intent that states should be able to continue these programs unencumbered with additional Federal requirements if they demonstrate that they meet the requirements of the Act. (U.S. House of Representatives, Committee on Interstate and Foreign Commerce, *Safe Drinking Water Act Amendments*, H. Rept. 96-1348 to accompany H.R. 8117, 96th Congress, 2d Session, September 19, 1980, p. 5.)

[30] SDWA § 1425 requires a state to demonstrate that its UIC program meets the requirements of § 1421(b)(1)(A) through (D) and represents an effective program (including adequate record keeping and reporting) to prevent underground injection which endangers underground sources of drinking water. To receive approval under § 1425's optional demonstration provisions, a state program must include permitting, inspection, monitoring, and record-keeping and reporting requirements.

[31] 42 U.S.C. § 300i. The Administrator may take action when information is received that (1) a contaminant is present in or is likely to enter a public drinking water supply system or underground source of drinking water "which may present an imminent and substantial endangerment to the health of persons," and (2) the appropriate state or local officials have not taken adequate action to protect such persons.

[32] 42 U.S.C. § 300h(b)(1).

[33] EPA further explained this requirement in a 1993 memorandum which provided that "[t]o better quantify the definition of USDW, EPA determined that any aquifer yielding more than 1 gallon per minute can be expected to provide sufficient quantity of water to serve a public water system and therefore falls under the definition of a USDW." EPA

Memorandum: *Assistance on Compliance of 40 CFR Part 191 with Ground Water Protection Standards*. From James R. Elder, Director, Office of Ground Water and Drinking Water, to Margo T. Oge, Director, Office of Radiation and Indoor Air, June 4, 1993.

[34] § 40 CFR 144.3. According to EPA regulations, an exempted aquifer is an aquifer, or a portion of an aquifer, that meets the criteria for a USDW, for which protection has been waived under the UIC program. Under 40 CFR Part 146.4, an aquifer may be exempted if it is not currently being used—and will not be used in the future—as a drinking water source, or it is not reasonably expected to supply a public water system due to a high total dissolved solids content. The SDWA does not mention aquifer exemption, but EPA explains that without aquifer exemptions, certain types of energy production, mining, or waste disposal into USDWs would be prohibited. EPA, typically at the Region level, makes the final determination on granting all exemptions.

[35] U.S. Environmental Protection Agency, *Evaluation of Impacts to Underground Sources of Drinking Water by Hydraulic Fracturing of Coalbed Methane Reservoirs*, EPA 816-R-04-003, June 2004, pp. 1-5.

[36] Regulatory requirements for state UIC programs are established in 40 CFR §§ 144-147.

[37] U.S. Environmental Protection Agency, Underground Injection Control Program, Classes of Wells, http://water

[38] Regulatory requirements for wells related to oil and gas production (Class II wells) are located at 40 CFR Parts 144 and 146.

[39] EPA historically has differentiated Class II wells from production wells. The agency's UIC website states that, "[p]roduction wells bring oil and gas to the surface; the UIC Program did not regulate production wells." U.S. Environmental Protection Agency, Class II Wells—Oil and Gas Related Injection Wells (Class II), "What are the types of Class II wells?," http://water

[40] To receive primacy, a state, territory, or Indian tribe must demonstrate to EPA that its UIC program is at least as stringent as the federal standards; the state, territory, or tribal UIC requirements may be more stringent than the federal requirements. For Class II wells, states must demonstrate that their programs are effective in preventing endangerment of underground sources of drinking water (USDWs). Requirements for state UIC programs are established in 40 CFR §§ 144-147.

[41] U.S. Environmental Protection Agency, *Guidance for State Submissions under Section 1425 of the Safe Drinking Water Act*, Ground Water Program Guidance #19, p. 20, http://www.epa.gov/safewater/uic/pdfs/guidance guide_uic_guidance-19_primacy_app.pdf.

[42] The House report for the 1980 Safe Drinking Water Act Amendments, H.R. 8117, which established § 1425, states that "So long as the statutory requirements are met, the states are not obligated to show that their programs mirror either procedurally or substantively the Administrator's regulations." H. Report to accompany H.R. 8117, No. 96-1348, September 19, 1980, p. 5.

[43] Legal Environmental Assistance Foundation, Inc. v. U.S. Environmental Protection Agency, 118F.3d 1467, 1471 (11th Cir. 1997) ("*LEAF I*").

[44] *Id.* at 1470.
[45] *Id.* at 1471.
[46] *Id.*
[47] *Id.* at 1472,
[48] *Id.* at 1473-74.
[49] *Id.* at 1474-75.
[50] *Id.* at 1478.

[51] *See* 64 Fed. Reg. 56986 (October 22, 1999).

[52] 42 U.S.C. § 300h-4(a).

[53] *Id.* at § 300h-1(b)(1)(A).

[54] Legal Environmental Assistance Foundation, Inc. v. U.S. Environmental Protection Agency, 276 F.3d 1253, 1257 (11th Cir. 2001). (*LEAF II*)

[55] 65 Fed. Reg. 2889 (October 2000).

[56] *Id.* at 1256.

[57] *Id.* at 1259-61.

[58] *Id.* at 1256.

[59] *Id.* at 1262.

[60] *Id.* at 1263.

[61] *Id.* at 1263-64.

[62] *Id.* at 1256 (referring to 5 U,.S.C. § 706(2)(A).

[63] *Id.* at 1265.

[64] Ala. Admin. Code, r. 400-3-8-.03(4), (2002). Responding to EPAct 2005 (see below), the state made some revisions to its regulations for hydraulic fracturing of coal beds in 2007. Ala. Admin. Code r. 400-3-8-.03(1).

[65] Ala. Admin. Code r. 400-3-8-.03(1).

[66] S. Marvin Rogers, State Oil and Gas Board of Alabama and Chairman, IOGCC Legal and Regulatory Affairs Committee, *History of Litigation Concerning Hydraulic Fracturing to Produce Coalbed Methane*, January 2009, p. 5.

[67] Ala. Admin. Code r. 400-3-8-.03(6)(a), 2002. To mitigate its increased administrative costs associated with implementation of the added regulations, operators pay a fee of $175 for each coalbed group fractured.

[68] A representative of the Alabama Coalbed Methane Association noted that the costs of hydraulic fracturing are very site specific and vary with operators as well as geology.

[69] Currently, EPA has authority to regulate only the use of diesel fuel in fracturing operations.

[70] U.S. Environmental Protection Agency. Draft Evaluation of Impacts to Underground Sources of Drinking Water by Hydraulic Fracturing of Coalbed Methane Reservoirs. EOA 816-D-02-006, August 2002.

[71] *Id.*, p. 6-20 - 6-21.

[72] National Drinking Water Advisory Council. Report on Hydraulic Fracturing and Underground Injection Control and Coalbed Methane by the National Drinking Water Advisory Council Resulting from a Conference Call Meeting Held December 12, 2002. Washington DC.

[73] U.S. Environmental Protection Agency, *Evaluation of Impacts to Underground Sources of Drinking Water by Hydraulic Fracturing of Coalbed Methane Reservoirs*, Final Report, EPA-816-04-003, Washington, D.C., June 2004, p. 4-1.

[74] *Evaluation of Impacts to USDWs by Hydraulic Fracturing of Coalbed Methane Reservoirs*, Final Report, p. 4-19.

[75] *Id.* p. 4-12.

[76] *Memorandum of Agreement Between the United States Environmental Protection Agency and BJ Services Company, Halliburton Energy Services, Inc., and Schlumberger Technology Corporation*, December 12, 2003.

[77] *LEAF I*, 118 F.3d at 1475

[78] http://water.epa.gov/type/groundwater/uic/class2/hydraulicfracturing/ wells_hydroreg.cfm# safehyfr,. The website notes that "[a]ny service company that performs hydraulic fracturing using diesel fuel must receive prior authorization from the UIC program," and that

"[i]njection wells receiving diesel fuel as a hydraulic fracturing additive will be considered Class II wells by the UIC program."

[79] Independent Petroleum Association of America and U.S. Oil & Gas Association v. U.S. Environmental Protection Agency, *Petition for Review*, D.C. Cir. Case No. 10-1233 (August 12, 2010).

[80] H.R. 1084, S. 587.

[81] H.R. 1084, at § 2(a). S. 587 is similar but does not include geothermal production activities.

[82] *Id.* at § 2(b).

[83] *Id.*

[84] In 2008, for example, the Colorado Oil and Gas Conservation Commission promulgated regulations requiring operators to maintain inventories of chemicals stored onsite for use downhole, and to provide a list of the chemicals of "trade secret chemical products" to commission officials upon request. Operators are also required to disclose chemical information to treating medical professionals. (2 Colo. Code Regs. § 404-1:205). Wyoming is another state that did not wait for the federal government to adopt disclosure requirements for persons engaged in hydraulic fracturing. On September 15, 2010, the Wyoming Oil and Gas Conservation Commission (WOGCC) promulgated its own set of hydraulic fracturing disclosure requirements. In accordance with these regulations, drilling operators are required to:

- identify all water supply wells within one-quarter mile of the drilling activity as well as the depth from which water is being appropriated (Wyo. Rules and Regs. Oil Gen § 3-8);
- provide stimulation fluid information to the WOGCC on its Application for Permit to Drill, as part of a comprehensive drilling/completion/recompletion plan, or on a separate notice (Wyo. Rules and Regs. Oil Gen § 3-45(a));
- provide geological names, geological description and depth of the formation into which well stimulation fluids are to be injected (Wyo. Rules and Regs. Oil Gen § 3-45(c));
- provide to an WOGCC Supervisor, for each stage of the well stimulation program, the chemical additives, compounds and concentrations or rates proposed to be mixed and injected, including (i) stimulation fluid identified by additive type; (ii) the chemical compound name and Chemical Abstracts Service (CAS) number of any constituents; and (iii) the proposed rate or concentration for each additive. The WOGCC Supervisor is also authorized to request additional information as deemed appropriate (Wyo. Rules and Regs. Oil Gen § 3-45(d)).
- provide a detailed description of the proposed well stimulation design, which shall include (i) the anticipated surface treating pressure range; (ii) the maximum injection treating pressure; and (iii) the estimated or calculated fracture length and fracture height.

The regulations prohibit the underground injection of "volatile organic compounds, such as benzene, toluene, ethylbenzene and xylene, also known as BTEX compounds or any petroleum distillates, into groundwater." (Wyo. Rules and Regs. Oil Gen. § 3-45(g)). The regulations do state that confidentiality protection will be provided for "trade secrets, privileged information and confidential commercial, financial, geological or geophysical data furnished by or obtained from any person." (Wyo. Rules and Regs. Oil Gen. § 3-45(f)). There also are logging requirements applicable to post-well stimulation (Wyo. Rules and Regs. Oil Gen. § 3-45(h)).

[85] H.R. 1084, § 2(b).

[86] The Occupational Safety and Health Administration has promulgated a set of regulations under Occupational Safety and Health Act (OSHAct), referred to as the Hazard Communication Standard (29 C.F.R. § 1910.1200). Additionally, OSHAct regulations require operators to maintain Material Safety Data Sheets (MSDS) for hazardous chemicals at the job site. The federal Emergency Planning and Community Right to Know Act (EPCRA) requires that facility owners submit an MSDS for each hazardous chemical present that exceeds an EPA-determined threshold level, or a list of such chemicals, to the local emergency planning committee (LEPC), the state emergency response commission, and the local fire department. For non-proprietary information, EPCRA generally requires a LEPC to provide an MSDS to a member of the public on request.

[87] The GWPC and Interstate Oil and Gas Compact Commission are launching a public registry of chemicals used in fracturing, with companies voluntarily identifying chemicals used in hydraulic fracturing in individual wells.

[88] U.S. Environmental Protection Agency, Region 8, *Hydraulic Fracturing*, Presentation, Underground Injection Control Program Meeting, Glenwood Springs, Colorado, August 8, 2009.

[89] § 1449; 42 U.S.C. 300j-8.

[90] In the case concerning Alabama, the Eleventh Circuit Court of Appeals ruled that "EPA's decision to subject hydraulic fracturing to approval under § 1425 rests upon a permissible construction of the Safe Drinking Water Act." *Legal Environmental Assistance Fund v. Environmental Protection Agency, State Oil and Gas Board of Alabama*, 276 F.3d 1253 (11th Cir. 2001).

[91] § 1421(b)(3)(A); 42 U.S.C. 300h(b)(3)(A).

[92] Because coal beds frequently are sources of drinking water, the Alabama State Oil and Gas Board requires well operators to certify that a proposed hydraulic fracturing operation would not occur in a USDW, or that the mixture of fracturing fluids would meet EPA drinking water standards. The state regulations also prohibit fracturing at depths shallower than 399 feet, as most drinking water wells rely on shallow aquifers.

[93] U.S. Department of Energy Office of Fossil Energy and National Technology Laboratory, *Modern Shale Gas Development in the United States: A Primer*, DE-FG26-04NT15455, April 2009, http://fossil.energy.gov/programs/oilgas/publications/ naturalgas_general/ Shale_Gas_Primer_2009.pdf.

[94] IHS Global Insight, *Measuring the Economic and Energy Impacts of Proposals to Regulate Hydraulic Fracturing*, Task 1 Report, Prepared for the American Petroleum Institute, Lexington, MA, 2009, p. 7.

[95] Advanced Resources International, Inc., *Potential Economic and Energy Supply Impacts of Proposals to Modify Federal Environmental Laws Applicable to the U.S. Oil and Gas Exploration and Production Industry*, U.S. Department of Energy, Office of Fossil Energy, January 2009. The authors note that cost estimates are based on a 1999 memorandum prepared for DOE, from Robin Petrusak, ICF Consulting to Nancy Johnson, U.S. Department of Energy, "Documentation of Estimated Potential Cost of Compliance for Toxic Release Inventory (TRI) Reporting and Hydraulic Fracturing," August 19, 1999.

[96] *Id.* p. 25-26.

[97] The GWPC is a national association representing state groundwater and UIC agencies whose mission is to promote protection and conservation of groundwater resources for beneficial uses. The stated purpose of the GWPC is "to promote and ensure the use of best

management practices and fair but effective laws regarding comprehensive ground water protection." http://www.gwpc.org/about_us/about_us.htm.

[98] Statement of Scott Kell, for the Ground Water Protection Council, House Committee on Natural Resources, Subcommittee on Energy and Mineral Resources, Oversight Hearing on "Unconventional Fuels, Part I: Shale Gas Potential," June 4, 2009.

[99] Scott Stevens and Vello Kuuskraa, Advanced Resources International, Inc., "Gas Shale-1: Seven Plays Dominate North America Activity," *Oil & Gas Journal*, vol. 107, no. 36 (September 28, 2009), p. 41.

[100] For a review of the applicability of various federal environmental laws to oil and gas development, see Amy Mall, Natural Resources Defense Council, "The applicability of federal requirements that protect public health and the environment to oil and gas development," Testimony before the Committee on Oversight and Government Reform, U.S. House of Representatives, October 31, 2007, http://www.nrdc.org/energy

[101] For information on efforts by the New York State Department of Environmental Conservation to develop regulations for unconventional gas reservoir development, see http://www.dec.ny.gov/energy

[102] The Interstate Oil and Gas Compact Commission represents the state oil and gas agencies. The commission was established in the 1930s, initially to reduce the waste of oil during exploration and production by developing model statutes and practices to improve the conservation of oil resources.

[103] The GWPC passed a resolution in 2003 encouraging Congress to clarify the definition of underground injection in Part C of SDWA to exclude the practice of hydraulic fracturing. http://www.gwpc.org/advocacyresolutions/RES-03-5.htm.

[104] Further policy positions and information can be found at the IOGCC website: http://www.ogcc.org/hydraulicfracturing.

[105] Hydraulic fracturing is used commonly used for conventional gas production. Wyoming, for example, reported that in 2008, 100% (1,316) of new conventional gas wells were fracture stimulated, many wells with multi-zone stimulations in each well bore, some staged, and some individual fracture stimulations. Source: Wyoming Oil and Gas Conservation Commission. The Commission rules require operators to receive approval prior to hydraulic fracturing treatments. Operators are required to provide detailed information regarding the fracturing process, including the source of water and/or trade name fluids, type of proppants, and estimated pump pressures. After a treatment is complete, the operator must provide fracturing data and production results.

[106] In October 2007, in response to the 2005 Energy Policy Act, Alabama revised its Class II UIC program to exclude hydraulic fracturing. The state retains most hydraulic fracturing requirements which it administers under its oil and gas regulatory regime.

[107] Ground Water Protection Council, U.S. Department of Energy, Office of Fossil Energy, National Energy Technology Laboratory, *State Oil and Natural Gas Regulations Designed to Protect Water Resources*, May 2009.

[108] More details of state rules are included in the Regulations Reference Document accompanying the GWPC report, *State Oil and Natural Gas Regulations Designed to Protect Water Resources*, http://www.gwpc.org/e-library/e_library_list.htm.

[109] Four states reported to GWPC that agencies other than the oil and gas authority are involved in the permit review process, either by requirement or upon request of the oil and gas agency. In 2008, Colorado revised its oil and gas regulations to allow for greater public participation in permitting and environmental assessment of oil and gas field sites. This expanded

participation includes review by other state water protection agencies. Ground Water Protection Council (2009).

[110] STRONGER, Inc., State Review of Oil and Natural Gas Environmental Regulations, Inc., http://www.strongerinc.org. The STRONGER state review process involves teams representing industry, states, environmental and public interest groups reviewing state oil and gas waste management programs.

[111] Ground Water Protection Council, *State Oil and Natural Gas Regulations Designed to Protect Water Resources*, pp. 7, 39-40. STRONGER has been reviewing state programs to assess whether any revisions may be needed to address issues surrounding the use of hydraulic fracturing in oil and gas development. Hydraulic fracturing reviews have been done recently for Pennsylvania, Ohio, Louisiana, and Oklahoma.

[112] New York has imposed a temporary moratorium on unconventional gas drilling until the state can update oil and gas regulations, specifically to address development of the Marcellus Shale and other tight shale formations in the state. See, New York State Department of Environmental Conservation and Division of Mineral Resources, *Draft Supplemental Generic Environmental Impact Statement on the Oil, Gas and Solution Mining Regulatory Program: Well Permit Issuance for Horizontal Drilling and High-Volume Hydraulic Fracturing to Develop the Marcellus Shale and Other Low-Permeability Gas Reservoirs*, September 2009, ftp://ftp.dec.state.ny.us/dmn/download/OGdSGEISFull.pdf.

[113] U.S. Environmental Protection Agency, FY2008 Drinking Water Factoids, EPA 816-K-08-004, November, 2008, http://www.epa.gov/safewater/databases/pdfs/ data_factoids_2008.pdf.

[114] EIA, *Natural Gas Navigator: Number of Producing Gas Wells*, August 2009, http://tonto.eia.doe.gov/dnav/ng/ngjrod_wells_s1_a.htm.

[115] Congress provided $10.9 million for each of FY2009 and FY2010.

[116] Mike Nickolaus, Ground Water Protection Council, UIC Funding Presentation, Ground Water Protection Council 2007 Meeting January 23, 2007, http://www.gwpc. org/ meetings/uic/2007/proceedings/Nickolaus_UIC07.pdf.

[117] Ground Water Protection Council, *Ground Water Report to the Nation: A Call to Action*, Underground Injection Control, Ch. 9, Oklahoma City, OK, 2007, http://www.gwpc.org. This estimate preceded EPA's promulgation of new UIC regulations establishing Class VI wells for geologic sequestration of carbon dioxide and EPA's determination that production wells that use diesel must receive a Class II permit.

[118] See requirements at, for example, 40 CFR 144.51(m), *Requirements prior to commencing injection*. Also, 40 CFR § 124.11 provides for public comments and requests for public hearings for UIC permits. The UIC program director is required to hold a public hearing whenever he or she finds a significant degree of public interest in a draft permit (40 CFR § 124.12(a)). § 124.13 states that a comment period may need to be longer than 30 days to allow commenters time to prepare and submit comments.

[119] Mike Nickolaus, Ground Water Protection Council, UIC Funding Presentation, January 23, 2007.

[120] USGS Fact Sheet at 5.

[121] U.S. Department of Energy, *DOE Projects to Advance Environmental Science and Technology: Nine Unconventional Natural Gas Projects Address Water Resource and Management Issues*, August 19, 2009. List of projects is available at http://www. fossil. energy

[122] P.L. 111-88, H.Rept. 111-316: Hydraulic Fracturing Study.—The conferees urge the Agency to carry out a study on the relationship between hydraulic fracturing and drinking water, using a credible approach that relies on the best available science, as well as independent sources of information. The conferees expect the study to be conducted through a transparent, peer-reviewed process that will ensure the validity and accuracy of the data. The Agency shall consult with other Federal agencies as well as appropriate State and interstate regulatory agencies in carrying out the study, which should be prepared in accordance with the Agency's quality assurance principles.

[123] Information on EPA's hydraulic fracturing study is available at http://water hydraulicfracturing/index.cfm.

In: Hydraulic Fracturing and Natural Gas Drilling   ISBN: 978-1-61470-180-4
Editor: Aarik Schultz   © 2012 Nova Science Publishers, Inc.

Chapter 2

CHEMICALS USED IN HYDRAULIC FRACTURING[*]

United States House of Representatives Committee on Energy and Commerce Minority Staff

PREPARED BY COMMITTEE STAFF FOR:
Henry A. Waxman
Ranking Member
Committee on Energy
and Commerce

Edward J. Markey
Ranking Member
Committee on Natural
Resources

Diana DeGette
Ranking Member
Subcommittee on Oversight
and Investigations

---

[*] This is an edited, reformatted and augmented version of a United States House of Representatives Committee on Energy and Commerce publication, dated APRIL 2011.

I. EXECUTIVE SUMMARY

Hydraulic fracturing has helped to expand natural gas production in the United States, unlocking large natural gas supplies in shale and other unconventional formations across the country. As a result of hydraulic fracturing and advances in horizontal drilling technology, natural gas production in 2010 reached the highest level in decades. According to new estimates by the Energy Information Administration (EIA), the United States possesses natural gas resources sufficient to supply the United States for approximately 110 years.

As the use of hydraulic fracturing has grown, so have concerns about its environmental and public health impacts. One concern is that hydraulic fracturing fluids used to fracture rock formations contain numerous chemicals that could harm human health and the environment, especially if they enter drinking water supplies. The opposition of many oil and gas companies to public disclosure of the chemicals they use has compounded this concern.

Last Congress, the Committee on Energy and Commerce launched an investigation to examine the practice of hydraulic fracturing in the United States. As part of that inquiry, the Committee asked the 14 leading oil and gas service companies to disclose the types and volumes of the hydraulic fracturing products they used in their fluids between 2005 and 2009 and the chemical contents of those products. This report summarizes the information provided to the Committee.

Between 2005 and 2009, the 14 oil and gas service companies used more than 2,500 hydraulic fracturing products containing 750 chemicals and other components. Overall, these companies used 780 million gallons of hydraulic fracturing products – not including water added at the well site – between 2005 and 2009.

Some of the components used in the hydraulic fracturing products were common and generally harmless, such as salt and citric acid. Some were unexpected, such as instant coffee and walnut hulls. And some were extremely toxic, such as benzene and lead. Appendix A lists each of the 750 chemicals and other components used in hydraulic fracturing products between 2005 and 2009.

The most widely used chemical in hydraulic fracturing during this time period, as measured by the number of compounds containing the chemical, was methanol. Methanol, which was used in 342 hydraulic fracturing products, is a hazardous air pollutant and is on the candidate list for potential regulation under the Safe Drinking Water Act. Some of the other most widely used chemicals were isopropyl alcohol (used in 274 products), 2-butoxyethanol (used in 126 products), and ethylene glycol (used in 119 products).

Between 2005 and 2009, the oil and gas service companies used hydraulic fracturing products containing 29 chemicals that are (1) known or possible human carcinogens, (2) regulated under the Safe Drinking Water Act for their risks to human health, or (3) listed as hazardous air pollutants under the Clean Air Act. These 29 chemicals were components of more than 650 different products used in hydraulic fracturing.

The BTEX compounds – benzene, toluene, xylene, and ethylbenzene – appeared in 60 of the hydraulic fracturing products used between 2005 and 2009. Each BTEX compound is a regulated contaminant under the Safe Drinking Water Act and a hazardous air pollutant under the Clean Air Act. Benzene also is a known human carcinogen. The hydraulic fracturing companies injected 11.4 million gallons of products containing at least one BTEX chemical over the five year period.

In many instances, the oil and gas service companies were unable to provide the Committee with a complete chemical makeup of the hydraulic fracturing fluids they used. Between 2005 and 2009, the companies used 94 million gallons of 279 products that contained at least one chemical or component that the manufacturers deemed proprietary or a trade secret. Committee staff requested that these companies disclose this proprietary information. Although some companies did provide information about these proprietary fluids, in most cases the companies stated that they did not have access to proprietary information about products they purchased "off the shelf" from chemical suppliers. In these cases, the companies are injecting fluids containing chemicals that they themselves cannot identify.

II. BACKGROUND

Hydraulic fracturing – a method by which oil and gas service companies provide access to domestic energy trapped in hard-to-reach geologic formations — has been the subject of both enthusiasm and increasing environmental and health concerns in recent years. Hydraulic fracturing, used in combination with horizontal drilling, has allowed industry to access natural gas reserves previously considered uneconomical, particularly in shale formations. As a result of the growing use of hydraulic fracturing, natural gas production in the United States reached 21,577 billion cubic feet in 2010, a level not achieved since a period of high natural gas production between 1970 and 1974.[1] Overall, the Energy Information Administration now projects that the United States possesses 2,552 trillion cubic feet of potential natural gas resources, enough to supply the United States for approximately 110 years. Natural gas from shale resources accounts for 827 trillion cubic feet of this total, which is more than double what the EIA estimated just a year ago.[2]

Hydraulic fracturing creates access to more natural gas supplies, but the process requires the use of large quantities of water and fracturing fluids, which are injected underground at high volumes and pressure. Oil and gas service companies design fracturing fluids to create fractures and transport sand or other granular substances to prop open the fractures. The composition of these fluids varies by formation, ranging from a simple mixture of water and sand to more complex mixtures with a multitude of chemical additives. The companies may use these chemical additives to thicken or thin the fluids, improve the flow of the fluid, or kill bacteria that can reduce fracturing performance.[3]

Some of these chemicals, if not disposed of safely or allowed to leach into the drinking water supply, could damage the environment or pose a risk to human health. During hydraulic fracturing, fluids containing chemicals are injected deep underground, where their migration is not entirely predictable. Well failures, such as the use of insufficient well casing, could lead to their release at shallower depths, closer to drinking water supplies.[4] Although some fracturing fluids are removed from the well at the end of the fracturing process, a substantial amount remains underground.[5]

While most underground injections of chemicals are subject to the protections of the Safe Drinking Water Act (SDWA), Congress in 2005 modified the law to exclude "the underground injection of fluids or propping agents (other than diesel fuels) pursuant to hydraulic fracturing operations related to oil, gas, or geothermal production activities" from the Act's protections.[6] Unless oil and gas service companies use diesel in the hydraulic fracturing process, the permanent underground injection of chemicals used for hydraulic fracturing is not regulated by the Environmental Protection Agency (EPA).

Concerns also have been raised about the ultimate outcome of chemicals that are recovered and disposed of as wastewater. This wastewater is stored in tanks or pits at the well site, where spills are possible.[7] For final disposal, well operators must either recycle the fluids for use in future fracturing jobs, inject it into underground storage wells (which, unlike the fracturing process itself, are subject to the Safe Drinking Water Act), discharge it to nearby surface water, or transport it to wastewater treatment facilities.[8] A recent report in the

New York Times raised questions about the safety of surface water discharge and the ability of water treatment facilities to process wastewater from natural gas drilling operations.[9]

Any risk to the environment and human health posed by fracturing fluids depends in large part on their contents. Federal law, however, contains no public disclosure requirements for oil and gas producers or service companies involved in hydraulic fracturing, and state disclosure requirements vary greatly.[10] While the industry has recently announced that it soon will create a public database of fluid components, reporting to this database is strictly voluntary, disclosure will not include the chemical identity of products labeled as proprietary, and there is no way to determine if companies are accurately reporting information for all wells.[11]

The absence of a minimum national baseline for disclosure of fluids injected during the hydraulic fracturing process and the exemption of most hydraulic fracturing injections from regulation under the Safe Drinking Water Act has left an informational void concerning the contents, chemical concentrations, and volumes of fluids that go into the ground during fracturing operations and return to the surface in the form of wastewater. As a result, regulators and the public are unable effectively to assess any impact the use of these fluids may have on the environment or public health.

III. METHODOLOGY

On February 18, 2010, the Committee commenced an investigation into the practice of hydraulic fracturing and its potential impact on water quality across the United States. This investigation built on work begun by Ranking Member Henry A. Waxman in 2007 as Chairman of the Committee on Oversight and Government Reform. The Committee initially sent letters to eight oil and gas service companies engaged in hydraulic fracturing in the United States. In May 2010, the Committee sent letters to six additional oil and gas service companies to assess a broader range of industry practices.[12] The February and May letters requested information on the type and volume of chemicals present in the hydraulic fracturing products that each company used in their fluids between 2005 and 2009.

The 14 oil and gas service companies that received the letter voluntarily provided substantial information to the Committee. As requested, the companies reported the names and volumes of the products they used during the five-year period.[13] For each hydraulic fracturing product reported, the companies also provided a Material Safety Data Sheet (MSDS) detailing the product's chemical components. The Occupational Safety and Health Administration (OSHA) requires chemical manufacturers to create a MSDS for every product they sell as a means to communicate potential health and safety hazards to employees and employers. The MSDS must list all hazardous ingredients if they comprise at least 1% of the product; for carcinogens, the reporting threshold is 0.1%.[14]

Under OSHA regulations, manufacturers may withhold the identity of chemical components that constitute "trade secrets."[15] If the MSDS for a particular product used by a company subject to the Committee's investigation reported that the identity of any chemical component was a trade secret, the Committee asked the company that used that product to provide the proprietary information, if available.

IV. HYDRAULIC FRACTURING FLUIDS AND THEIR CONTENTS

Between 2005 and 2009, the 14 oil and gas service companies used more than 2,500 hydraulic fracturing products containing 750 chemicals and other components.[16] Overall, these companies used 780 million gallons of hydraulic fracturing products in their fluids between 2005 and 2009. This volume does not include water that the companies added to the fluids at the well site before injection. The products are comprised of a wide range of chemicals. Some are seemingly harmless like sodium chloride (salt), gelatin, and citric acid. Others could pose a severe risk to human health or the environment.

Some of the components were surprising. One company told the Committee that it used instant coffee as one of the components in a fluid designed to inhibit acid corrosion. Two companies reported using walnut hulls as part of a breaker—a product used to degrade the fracturing fluid viscosity, which helps to enhance post-fracturing fluid recovery. Another company reported using carbohydrates as a breaker. One company used tallow soap— soap made from beef, sheep, or other animals—to reduce loss of fracturing fluid into the exposed rock.

Appendix A lists each of the 750 chemicals and other components used in the hydraulic fracturing products injected underground between 2005 and 2009.

A. Commonly Used Chemical Components

The most widely used chemical in hydraulic fracturing during this time period, as measured by the number of products containing the chemical, was methanol. Methanol is a hazardous air pollutant and a candidate for regulation under the Safe Drinking Water Act. It was a component in 342 hydraulic fracturing products. Some of the other most widely used chemicals include isopropyl alcohol, which was used in 274 products, and ethylene glycol, which was used in 119 products. Crystalline silica (silicon dioxide) appeared in 207 products, generally proppants used to hold open fractures. Table 1 has a list of the most commonly used compounds in hydraulic fracturing fluids.

Table 1. Chemical Components Appearing Most Often in Hydraulic Fracturing Products Used Between 2005 and 2009

Chemical Component	No. of Products Containing Chemical
Methanol (Methyl alcohol)	342
Isopropanol (Isopropyl alcohol, Propan-2-ol)	274
Crystalline silica - quartz (SiO2)	207
Ethylene glycol monobutyl ether (2-	126
Ethylene glycol (1,2-ethanediol)	119
Hydrotreated light petroleum distillates	89
Sodium hydroxide (Caustic soda)	80

Hydraulic fracturing companies used 2-butoxyethanol (2-BE) as a foaming agent or surfactant in 126 products. According to EPA scientists, 2-BE is easily absorbed and rapidly distributed in humans following inhalation, ingestion, or dermal exposure. Studies have shown that exposure to 2-BE can cause hemolysis (destruction of red blood cells) and damage to the spleen, liver, and bone marrow.[17] The hydraulic fracturing companies injected 21.9 million gallons of products containing 2-BE between 2005 and 2009. They used the highest volume of products containing 2-BE in Texas, which accounted for more than half of the volume used. EPA recently found this chemical in drinking water wells tested in Pavillion, Wyoming.[18] Table 2 shows the use of 2-BE by state.

Table 2. States with the Highest Volume of Hydraulic Fracturing Fluids Containing 2-Butoxyethanol (2005-2009)

State	Fluid Volume (gallons)
Texas	12,031,734
Oklahoma	2,186,613
New Mexico	1,871,501
Colorado	1,147,614
Louisiana	890,068
Pennsylvania	747,416
West Virginia	464,231
Utah	382,874
Montana	362,497
Arkansas	348,959

B. Toxic Chemicals

The oil and gas service companies used hydraulic fracturing products containing 29 chemicals that are (1) known or possible human carcinogens, (2) regulated under the Safe Drinking Water Act for their risks to human health, or (3) listed as hazardous air pollutants under the Clean Air Act. These 29 chemicals were components of 652 different products used in hydraulic fracturing. Table 3 lists these toxic chemicals and their frequency of use.

Table 3. Chemicals Components of Concern: Carcinogens, SDWA-Regulated Chemicals, and Hazardous Air Pollutants

Chemical Component	Chemical Category	No. of Products
Methanol (Methyl alcohol)	HAP	342
Ethylene glycol (1,2-ethanediol)	HAP	119
Diesel[19]	Carcinogen, SDWA, HAP	51
Naphthalene	Carcinogen, HAP	44
Xylene	SDWA, HAP	44
Hydrogen chloride (Hydrochloric acid)	HAP	42
Toluene	SDWA, HAP	29

Chemical Component	Chemical Category	No. of Products
Ethylbenzene	SDWA, HAP	28
Diethanolamine (2,2-iminodiethanol)	HAP	14
Formaldehyde	Carcinogen, HAP	12
Sulfuric acid	Carcinogen	9
Thiourea	Carcinogen	9
Benzyl chloride	Carcinogen, HAP	8
Cumene	HAP	6
Nitrilotriacetic acid	Carcinogen	6
Dimethyl formamide	HAP	5
Phenol	HAP	5
Benzene	Carcinogen, SDWA, HAP	3
Di (2-ethylhexyl) phthalate	Carcinogen, SDWA, HAP	3
Acrylamide	Carcinogen, SDWA, HAP	2
Hydrogen fluoride (Hydrofluoric acid)	HAP	2
Phthalic anhydride	HAP	2
Acetaldehyde	Carcinogen, HAP	1
Acetophenone	HAP	1
Copper	SDWA	1
Ethylene oxide	Carcinogen, HAP	1
Lead	Carcinogen, SDWA, HAP	1
Propylene oxide	Carcinogen, HAP	1
p-Xylene	HAP	1
Number of Products Containing a Component of Concern		652

1. Carcinogens

Between 2005 and 2009, the hydraulic fracturing companies used 95 products containing 13 different carcinogens.[20] These included naphthalene (a possible human carcinogen), benzene (a known human carcinogen), and acrylamide (a probable human carcinogen). Overall, these companies injected 10.2 million gallons of fracturing products containing at least one carcinogen. The companies used the highest volume of fluids containing one or more carcinogens in Texas, Colorado, and Oklahoma. Table 4 shows the use of these chemicals by state.

Table 4. States with at Least 100,000 Gallons of Hydraulic Fracturing Fluids Containing a Carcinogen (2005-2009)

State	Fluid Volume (gallons)
Texas	3,877,273
Colorado	1,544,388
Oklahoma	1,098,746
Louisiana	777,945
Wyoming	759,898
North Dakota	557,519
New Mexico	511,186
Montana	394,873
Utah	382,338

2. Safe Drinking Water Act Chemicals

Under the Safe Drinking Water Act, EPA regulates 53 chemicals that may have an adverse effect on human health and are known to or likely to occur in public drinking water systems at levels of public health concern. Between 2005 and 2009, the hydraulic fracturing companies used 67 products containing at least one of eight SDWA-regulated chemicals.

Overall, they injected 11.7 million gallons of fracturing products containing at least one chemical regulated under SDWA. Most of these chemicals were injected in Texas. Table 5 shows the use of these chemicals by state.

The vast majority of these SDWA-regulated chemicals were the BTEX compounds – benzene, toluene, xylene, and ethylbenzene. The BTEX compounds appeared in 60 hydraulic fracturing products used between 2005 and 2009 and were used in 11.4 million gallons of hydraulic fracturing fluids. The Department of Health and Human Services, the International Agency for Research on Cancer, and EPA have determined that benzene is a human carcinogen.[21] Chronic exposure to toluene, ethylbenzene, or xylenes also can damage the central nervous system, liver, and kidneys.[22]

Table 5. States with at Least 100,000 Gallons of Hydraulic Fracturing Fluids Containing a SDWA-Regulated Chemical (2005-2009)

State	Fluid Volume (gallons)
Texas	9,474,631
New Mexico	1,157,721
Colorado	375,817
Oklahoma	202,562
Mississippi	108,809
North Dakota	100,479

In addition, the hydraulic fracturing companies injected more than 30 million gallons of diesel fuel or hydraulic fracturing fluids containing diesel fuel in wells in 19 states.[23] In a 2004 report, EPA stated that the "use of diesel fuel in fracturing fluids poses the greatest threat" to underground sources of drinking water.[24] Diesel fuel contains toxic constituents, including BTEX compounds.[25]

EPA also has created a Candidate Contaminant List (CCL), which is a list of contaminants that are currently not subject to national primary drinking water regulations but are known or anticipated to occur in public water systems and may require regulation under the Safe Drinking Water Act in the future.[26] Nine chemicals on that list—1-butanol, acetaldehyde, benzyl chloride, ethylene glycol, ethylene oxide, formaldehyde, methanol, n-methyl-2-pyrrolidone, and propylene oxide—were used in hydraulic fracturing products between 2005 and 2009.

3. Hazardous Air Pollutants

The Clean Air Act requires EPA to control the emission of 187 hazardous air pollutants, which are pollutants that cause or may cause cancer or other serious health effects, such as reproductive effects or birth defects, or adverse environmental and ecological effects.[27] Between 2005 and 2009, the hydraulic fracturing companies used 595 products containing 24 different hazardous air pollutants.

Hydrogen fluoride is a hazardous air pollutant that is a highly corrosive and systemic poison that causes severe and sometimes delayed health effects due to deep tissue penetration. Absorption of substantial amounts of hydrogen fluoride by any route may be fatal.[28] One of the hydraulic fracturing companies used 67,222 gallons of two products containing hydrogen fluoride in 2008 and 2009.

Lead is a hazardous air pollutant that is a heavy metal that is particularly harmful to children's neurological development. It also can cause health problems in adults, including reproductive problems, high blood pressure, and nerve disorders.[29] One of the hydraulic fracturing companies used 780 gallons of a product containing lead in this five-year period.

Methanol is the hazardous air pollutant that appeared most often in hydraulic fracturing products. Other hazardous air pollutants used in hydraulic fracturing fluids included formaldehyde, hydrogen chloride, and ethylene glycol.

V. USE OF PROPRIETARY AND "TRADE SECRET" CHEMICALS

Many chemical components of hydraulic fracturing fluids used by the companies were listed on the MSDSs as "proprietary" or "trade secret." The hydraulic fracturing companies used 93.6 million gallons of 279 products containing at least one proprietary component between 2005 and 2009.[30]

The Committee requested that these companies disclose this proprietary information. Although a few companies were able to provide additional information to the Committee about some of the fracturing products, in most cases the companies stated that they did not have access to proprietary information about products they purchased "off the shelf" from chemical suppliers. The proprietary information belongs to the suppliers, not the users of the chemicals.

Universal Well Services, for example, told the Committee that it "obtains hydraulic fracturing products from third-party manufacturers, and to the extent not publicly disclosed, product composition is proprietary to the respective vendor and not to the Company."[31] Complete Production Services noted that the company always uses fluids from third-party suppliers who provide an MSDS for each product. Complete confirmed that it is "not aware of any circumstances in which the vendors who provided the products have disclosed this proprietary information" to the company, further noting that "such information is highly proprietary for these vendors, and would not generally be disclosed to service providers" like Complete.[32] Key Energy Services similarly stated that it "generally does not have access to the trade secret information as a purchaser of the chemical(s)."[33] Trican also told the Committee that it has limited knowledge of "off the shelf" products purchased from a chemical distributor or manufacturer, noting that "Trican does not have any information in its possession about the components of such products beyond what the distributor of each product provided Trican in the MSDS sheet."[34]

In these cases, it appears that the companies are injecting fluids containing unknown chemicals about which they may have limited understanding of the potential risks posed to human health and the environment.

VI. CONCLUSION

Hydraulic fracturing has opened access to vast domestic reserves of natural gas that could provide an important stepping stone to a clean energy future. Yet questions about the safety of hydraulic fracturing persist, which are compounded by the secrecy surrounding the chemicals used in hydraulic fracturing fluids. This analysis is the most comprehensive national assessment to date of the types and volumes of chemical used in the hydraulic fracturing process. It shows that between 2005 and 2009, the 14 leading hydraulic fracturing companies in the United States used over 2,500 hydraulic fracturing products containing 750 compounds. More than 650 of these products contained chemicals that are known or possible human carcinogens, regulated under the Safe Drinking Water Act, or listed as hazardous air pollutants.

APPENDIX A. CHEMICAL COMPONENTS OF HYDRAULIC FRACTURING PRODUCTS, 2005-2009[35]

Chemical Component	Chemical Abstract Service Number	No. of Products Containing Chemical
1-(1-naphthylmethyl)quinolinium chloride	65322-65-8	1
1,2,3-propanetricarboxylic acid, 2-hydroxy-, trisodium salt, dihydrate	6132-04-3	1
1,2,3-trimethylbenzene	526-73-8	1
1,2,4-trimethylbenzene	95-63-6	21
1,2-benzisothiazol-3	2634-33-5	1

Appendix A. (Continued).

Chemical Component	Chemical Abstract Service Number	No. of Products Containing Chemical
1,2-dibromo-2,4-dicyanobutane	35691-65-7	1
1,2-ethanediaminium, N, N'-bis[2-[bis(2-hydroxyethyl)methylammonio]ethyl]-N,N'-bis(2-hydroxyethyl)-N,N'-dimethyl-,tetrachloride	138879-94-4	2
1,3,5-trimethylbenzene	108-67-8	3
1,6-hexanediamine dihydrochloride	6055-52-3	1
1,8-diamino-3,6-dioxaoctane	929-59-9	1
1-hexanol	111-27-3	1
1-methoxy-2-propanol	107-98-2	3
2,2`-azobis (2-amidopropane) dihydrochloride	2997-92-4	1
2,2-dibromo-3-nitrilopropionamide	10222-01-2	27
2-acrylamido-2-methylpropanesulphonic acid sodium salt polymer	*	1
2-bromo-2-nitropropane-1,3-diol	52-51-7	4
2-butanone oxime	96-29-7	1
2-hydroxypropionic acid	79-33-4	2
2-mercaptoethanol (Thioglycol)	60-24-2	13
2-methyl-4-isothiazolin-3-one	2682-20-4	4
2-monobromo-3-nitrilopropionamide	1113-55-9	1
2-phosphonobutane-1,2,4-tricarboxylic acid	37971-36-1	2
2-phosphonobutane-1,2,4-tricarboxylic acid, potassium salt	93858-78-7	1
2-substituted aromatic amine salt	*	1
4,4'-diaminodiphenyl sulfone	80-08-0	3
5-chloro-2-methyl-4-isothiazolin-3-one	26172-55-4	5
Acetaldehyde	75-07-0	1
Acetic acid	64-19-7	56
Acetic anhydride	108-24-7	7
Acetone	67-64-1	3
Acetophenone	98-86-2	1
Acetylenic alcohol	*	1

Chemicals Used in Hydraulic Fracturing

Chemical Component	Chemical Abstract Service Number	No. of Products Containing Chemical
Acetyltriethyl citrate	77-89-4	1
Acrylamide	79-06-1	2
Acrylamide copolymer	*	1
Acrylamide copolymer	38193-60-1	1
Acrylate copolymer	*	1
Acrylic acid, 2-hydroxyethyl ester	818-61-1	1
Acrylic acid/2-acrylamido-methylpropylsulfonic acid copolymer	37350-42-8	1
Acrylic copolymer	403730-32-5	1
Acrylic polymers	*	1
Acrylic polymers	26006-22-4	2
Acyclic hydrocarbon blend	*	1
Adipic acid	124-04-9	6
Alcohol alkoxylate	*	5
Alcohol ethoxylates	*	2
Alcohols	*	9
Alcohols, C11-15-secondary, ethoxylated	68131-40-8	1
Alcohols, C12-14-secondary	126950-60-5	4
Alcohols, C12-14-secondary, ethoxylated	84133-50-6	19
Alcohols, C12-15, ethoxylated	68131-39-5	2
Alcohols, C12-16, ethoxylated	103331-86-8	1
Alcohols, C12-16, ethoxylated	68551-12-2	3
Alcohols, C14-15, ethoxylated	68951-67-7	5
Alcohols, C9-11-iso-, C10-rich, ethoxylated	78330-20-8	4
Alcohols, C9-C22	*	1
Aldehyde	*	4
Aldol	107-89-1	1
Alfa-Alumina	*	5
Aliphatic acid	*	1
Aliphatic alcohol polyglycol ether	68015-67-8	1
Aliphatic amine derivative	120086-58-0	2
Alkaline bromide salts	*	2
Alkanes, C10-14	93924-07-3	2
Alkanes, C13-16-iso	68551-20-2	2

Appendix A. (Continued).

Chemical Component	Chemical Abstract Service Number	No. of Products Containing Chemical
Alkanolamine	150-25-4	3
Alkanolamine chelate of zirconium alkoxide (Zirconium complex)	197980-53-3	4
Alkanolamine/aldehyde condensate	*	1
Alkenes	*	1
Alkenes, C>10 alpha-	64743-02-8	3
Alkenes, C>8	68411-00-7	2
Alkoxylated alcohols	*	1
Alkoxylated amines	*	6
Alkoxylated phenol formaldehyde resin	63428-92-2	1
Alkyaryl sulfonate	*	1
Alkyl (C12-16) dimethyl benzyl ammonium chloride	68424-85-1	7
Alkyl (C6-C12) alcohol, ethoxylated	68439-45-2	2
Alkyl (C9-11) alcohol, ethoxylated	68439-46-3	1
Alkyl alkoxylate	*	9
Alkyl amine	*	2
Alkyl amine blend in a metal salt solution	*	1
Alkyl aryl amine sulfonate	255043-08-04	1
Alkyl benzenesulfonic acid	68584-22-5	2
Alkyl esters	*	2
Alkyl hexanol	*	1
Alkyl ortho phosphate ester	*	1
Alkyl phosphate ester	*	3
Alkyl quaternary ammonium chlorides	*	4
Alkylaryl sulfonate	*	1
Alkylaryl sulphonic acid	27176-93-9	1
Alkylated quaternary chloride	*	5
Alkylbenzenesulfonic acid	*	1
Alkylethoammonium sulfates	*	1
Alkylphenol ethoxylates	*	1
Almandite and pyrope garnet	1302-62-1	1
Aluminium isopropoxide	555-31-7	1
Aluminum	7429-90-5	2

Chemicals Used in Hydraulic Fracturing

Chemical Component	Chemical Abstract Service Number	No. of Products Containing Chemical
Aluminum chloride	*	3
Aluminum chloride	1327-41-9	2
Aluminum oxide (alpha-Alumina)	1344-28-1	24
Aluminum oxide silicate	12068-56-3	1
Aluminum silicate (mullite)	1302-76-7	38
Aluminum sulfate hydrate	10043-01-3	1
Amides, tallow, n-[3-	68647-77-8	4
Amidoamine	*	1
Amine	*	7
Amine bisulfite	13427-63-9	1
Amine oxides	*	1
Amine phosphonate	*	3
Amine salt	*	2
Amines, C14-18; C16-18-unsaturated, alkyl, ethoxylated	68155-39-5	1
Amines, coco alkyl, acetate	61790-57-6	3
Amines, polyethylenepoly-, ethoxylated, phosphonomethylated	68966-36-9	1
Amines, tallow alkyl, ethoxylated	61791-26-2	2
Amino compounds	*	1
Amino methylene phosphonic acid salt	*	1
Amino trimethylene phosphonic acid	6419-19-8	2
Ammonia	7664-41-7	7
Ammonium acetate	631-61-8	4
Ammonium alcohol ether sulfate	68037-05-8	1
Ammonium bicarbonate	1066-33-7	1
Ammonium bifluoride (Ammonium hydrogen difluoride)	1341-49-7	10
Ammonium bisulfate	7783-20-2	3
Ammonium bisulfite	10192-30-0	15
Ammonium C6-C10 alcohol ethoxysulfate	68187-17-7	4
Ammonium C8-C10 alkyl ether sulfate	68891-29-2	4
Ammonium chloride	12125-02-9	29
Ammonium fluoride	12125-01-8	9
Ammonium hydroxide	1336-21-6	4
Ammonium nitrate	6484-52-2	2

Appendix A. (Continued).

Chemical Component	Chemical Abstract Service Number	No. of Products Containing Chemical
Ammonium persulfate (Diammonium peroxidisulfate)	7727-54-0	37
Ammonium salt	*	1
Ammonium salt of ethoxylated alcohol sulfate	*	1
Amorphous silica	99439-28-8	1
Amphoteric alkyl amine	61789-39-7	1
Anionic copolymer	*	3
Anionic polyacrylamide	*	1
Anionic polyacrylamide	25085-02-3	6
Anionic polyacrylamide copolymer	*	3
Anionic polymer	*	2
Anionic polymer in solution	*	1
Anionic polymer, sodium salt	9003-04-7	1
Anionic water-soluble polymer	*	2
Antifoulant	*	1
Antimonate salt	*	1
Antimony pentoxide	1314-60-9	2
Antimony potassium oxide	29638-69-5	4
Antimony trichloride	10025-91-9	2
a-organic surfactants	61790-29-8	1
Aromatic alcohol glycol ether	*	2
Aromatic aldehyde	*	2
Aromatic ketones	224635-63-6	2
Aromatic polyglycol ether	*	1
Barium sulfate	7727-43-7	3
Bauxite	1318-16-7	16
Bentonite	1302-78-9	2
Benzene	71-43-2	3
Benzene, C10-16, alkyl derivatives	68648-87-3	1
Benzenecarboperoxoic acid, 1,1-dimethylethyl ester	614-45-9	1
Benzenemethanaminium	3844-45-9	1

Chemicals Used in Hydraulic Fracturing

Chemical Component	Chemical Abstract Service Number	No. of Products Containing Chemical
Benzenesulfonic acid, C10-16-alkyl derivs., potassium salts	68584-27-0	1
Benzoic acid	65-85-0	11
Benzyl chloride	100-44-7	8
Biocide component	*	3
Bis(1-methylethyl)naphthalenesulfonic acid, cyclohexylamine salt	68425-61-6	1
Bishexamethylenetriamine penta methylene phosphonic acid	35657-77-3	1
Bisphenol A/Epichlorohydrin resin	25068-38-6	5
Bisphenol A/Novolac epoxy resin	28906-96-9	1
Borate	12280-03-4	2
Borate salts	*	5
Boric acid	10043-35-3	18
Boric acid, potassium salt	20786-60-1	1
Boric acid, sodium salt	1333-73-9	2
Boric oxide	1303-86-2	1
b-tricalcium phosphate	7758-87-4	1
Butanedioic acid	2373-38-8	4
Butanol	71-36-3	3
Butyl glycidyl ether	2426-08-6	5
Butyl lactate	138-22-7	4
C10-C16 ethoxylated alcohol	68002-97-1	4
C-11 to C-14 n-alkanes, mixed	*	1
C12-C14 alcohol, ethoxylated	68439-50-9	3
Calcium carbonate	471-34-1	1
Calcium carbonate (Limestone)	1317-65-3	9
Calcium chloride	10043-52-4	17
Calcium chloride, dihydrate	10035-04-8	1
Calcium fluoride	7789-75-5	2
Calcium hydroxide	1305-62-0	9
Calcium hypochlorite	7778-54-3	1
Calcium oxide	1305-78-8	6
Calcium peroxide	1305-79-9	5
Carbohydrates	*	3
Carbon dioxide	124-38-9	4

Appendix A. (Continued).

Chemical Component	Chemical Abstract Service Number	No. of Products Containing Chemical
Carboxymethyl guar gum, sodium salt	39346-76-4	7
Carboxymethyl hydroxypropyl guar	68130-15-4	11
Cellophane	9005-81-6	2
Cellulase	9012-54-8	7
Cellulase enzyme	*	1
Cellulose	9004-34-6	1
Cellulose derivative	*	2
Chloromethylnaphthalene quinoline quaternary amine	15619-48-4	3
Chlorous ion solution	*	2
Choline chloride	67-48-1	3
Chromates	*	1
Chromium (iii) acetate	1066-30-4	1
Cinnamaldehyde (3-phenyl-2-propenal)	104-55-2	5
Citric acid (2-hydroxy-1,2,3	77-92-9	29
Citrus terpenes	94266-47-4	11
Coal, granular	50815-10-6	1
Cobalt acetate	71-48-7	1
Cocaidopropyl betaine	61789-40-0	2
Cocamidopropylamine oxide	68155-09-9	1
Coco bis-(2-hydroxyethyl) amine oxide	61791-47-7	1
Cocoamidopropyl betaine	70851-07-9	1
Cocomidopropyl dimethylamine	68140-01-2	1
Coconut fatty acid diethanolamide	68603-42-9	1
Collagen (Gelatin)	9000-70-8	6
Complex alkylaryl polyo-ester	*	1
Complex aluminum salt	*	2
Complex organometallic salt	*	2
Complex substituted keto-amine	143106-84-7	1
Complex substituted keto-amine	*	1
Copolymer of acrylamide and sodium acrylate	25987-30-8	1
Copper	7440-50-8	1

Chemicals Used in Hydraulic Fracturing

Chemical Component	Chemical Abstract Service Number	No. of Products Containing Chemical
Copper iodide	7681-65-4	1
Copper sulfate	7758-98-7	3
Corundum (Aluminum oxide)	1302-74-5	48
Crotonaldehyde	123-73-9	1
Crystalline silica - cristobalite	14464-46-1	44
Crystalline silica - quartz (SiO_2)	14808-60-7	207
Crystalline silica, tridymite	15468-32-3	2
Cumene	98-82-8	6
Cupric chloride	7447-39-4	10
Cupric chloride dihydrate	10125-13-0	7
Cuprous chloride	7758-89-6	1
Cured acrylic resin	*	7
Cured resin	*	4
Cured silicone rubber-polydimethylsiloxane	63148-62-9	1
Cured urethane resin	*	3
Cyclic alkanes	*	1
Cyclohexane	110-82-7	1
Cyclohexanone	108-94-1	1
Decanol	112-30-1	2
Decyl-dimethyl amine oxide	2605-79-0	4
Dextrose monohydrate	50-99-7	1
D-Glucitol	50-70-4	1
Di (2-ethylhexyl) phthalate	117-81-7	3
Di (ethylene glycol) ethyl ether acetate	112-15-2	4
Diatomaceous earth	61790-53-2	3
Diatomaceous earth, calcined	91053-39-3	7
Dibromoacetonitrile	3252-43-5	1
Dibutylaminoethanol (2-	102-81-8	4
Di-calcium silicate	10034-77-2	1
Dicarboxylic acid	*	1
Didecyl dimethyl ammonium chloride	7173-51-5	1
Diesel	*	1
Diesel	68334-30-5	3
Diesel	68476-30-2	4
Diesel	68476-34-6	43

Appendix A. (Continued).

Chemical Component	Chemical Abstract Service Number	No. of Products Containing Chemical
Diethanolamine (2,2-iminodiethanol)	111-42-2	14
Diethylbenzene	25340-17-4	1
Diethylene glycol	111-46-6	8
Diethylene glycol monomethyl ether	111-77-3	4
Diethylene triaminepenta (methylene phosphonic acid)	15827-60-8	1
Diethylenetriamine	111-40-0	2
Diethylenetriamine, tall oil fatty acids	61790-69-0	1
Diisopropylnaphthalenesulfonic acid	28757-00-8	2
Dimethyl formamide	68-12-2	5
Dimethyl glutarate	1119-40-0	1
Dimethyl silicone	*	2
Dioctyl sodium sulfosuccinate	577-11-7	1
Dipropylene glycol	25265-71-8	1
Dipropylene glycol monomethyl ether (2-methoxymethylethoxy propanol)	34590-94-8	12
Di-secondary-butylphenol	53964-94-6	3
Disodium EDTA	139-33-3	1
Disodium ethylenediaminediacetate	38011-25-5	1
Disodium ethylenediaminetetraacetate	6381-92-6	1
Disodium octaborate tetrahydrate	12008-41-2	1
Dispersing agent	*	1
d-Limonene	5989-27-5	11
Dodecyl alcohol ammonium sulfate	32612-48-9	2
Dodecylbenzene sulfonic acid	27176-87-0	14
Dodecylbenzene sulfonic acid salts	42615-29-2	2
Dodecylbenzene sulfonic acid salts	68648-81-7	7
Dodecylbenzene sulfonic acid salts	90218-35-2	1
Dodecylbenzenesulfonate isopropanolamine	42504-46-1	1
Dodecylbenzenesulfonic acid, monoethanolamine salt	26836-07-7	1

Chemicals Used in Hydraulic Fracturing

Chemical Component	Chemical Abstract Service Number	No. of Products Containing Chemical
Dodecylbenzenesulphonic acid, morpholine salt	12068-08-5	1
EDTA/Copper chelate	*	2
EO-C7-9-iso-, C8-rich alcohols	78330-19-5	5
Epichlorohydrin	25085-99-8	5
Epoxy resin	*	5
Erucic amidopropyl dimethyl betaine	149879-98-1	3
Erythorbic acid	89-65-6	2
Essential oils	*	6
Ethanaminium, n,n,n-trimethyl-2-[(1-oxo-2-propenyl)oxy]-,chloride, polymer with 2-propenamide	69418-26-4	4
Ethanol (Ethyl alcohol)	64-17-5	36
Ethanol, 2-(hydroxymethylamino)-	34375-28-5	1
Ethanol, 2, 2'-(Octadecylamino) bis-	10213-78-2	1
Ethanoldiglycine disodium salt	135-37-5	1
Ether salt	25446-78-0	2
Ethoxylated 4-nonylphenol (Nonyl phenol ethoxylate)	26027-38-3	9
Ethoxylated alcohol	104780-82-7	1
Ethoxylated alcohol	78330-21-9	2
Ethoxylated alcohols	*	3
Ethoxylated alkyl amines	*	1
Ethoxylated amine	*	1
Ethoxylated amines	61791-44-4	1
Ethoxylated fatty acid ester	*	1
Ethoxylated nonionic surfactant	*	1
Ethoxylated nonyl phenol	*	8
Ethoxylated nonyl phenol	68412-54-4	10
Ethoxylated nonyl phenol	9016-45-9	38
Ethoxylated octyl phenol	68987-90-6	1
Ethoxylated octyl phenol	9002-93-1	1
Ethoxylated octyl phenol	9036-19-5	3
Ethoxylated oleyl amine	13127-82-7	2
Ethoxylated oleyl amine	26635-93-8	1

Appendix A. (Continued).

Chemical Component	Chemical Abstract Service Number	No. of Products Containing Chemical
Ethoxylated sorbitol esters	*	1
Ethoxylated tridecyl alcohol phosphate	9046-01-9	2
Ethoxylated undecyl alcohol	127036-24-2	2
Ethyl acetate	141-78-6	4
Ethyl acetoacetate	141-97-9	1
Ethyl octynol (1-octyn-3-ol,4-ethyl-)	5877-42-9	5
Ethylbenzene	100-41-4	28
Ethylene glycol (1,2-ethanediol)	107-21-1	119
Ethylene glycol monobutyl ether (2-butoxyethanol)	111-76-2	126
Ethylene oxide	75-21-8	1
Ethylene oxide-nonylphenol polymer	*	1
Ethylenediaminetetraacetic acid	60-00-4	1
Ethylene-vinyl acetate copolymer	24937-78-8	1
Ethylhexanol (2-ethylhexanol)	104-76-7	18
Fatty acid ester	*	1
Fatty acid, tall oil, hexa esters with sorbitol, ethoxylated	61790-90-7	1
Fatty acids	*	1
Fatty alcohol alkoxylate	*	1
Fatty alkyl amine salt	*	1
Fatty amine carboxylates	*	1
Fatty quaternary ammonium chloride	61789-68-2	1
Ferric chloride	7705-08-0	3
Ferric sulfate	10028-22-5	7
Ferrous sulfate, heptahydrate	7782-63-0	4
Fluoroaliphatic polymeric esters	*	1
Formaldehyde	50-00-0	12
Formaldehyde polymer	*	2
Formaldehyde, polymer with 4-(1,1-dimethyl)phenol, methyloxirane and oxirane	30704-64-4	3
Formaldehyde, polymer with 4-nonylphenol and oxirane	30846-35-6	1

Chemicals Used in Hydraulic Fracturing

Chemical Component	Chemical Abstract Service Number	No. of Products Containing Chemical
Formaldehyde, polymer with ammonia and phenol	35297-54-2	2
Formamide	75-12-7	5
Formic acid	64-18-6	24
Fumaric acid	110-17-8	8
Furfural	98-01-1	1
Furfuryl alcohol	98-00-0	3
Glass fiber	65997-17-3	3
Gluconic acid	526-95-4	1
Glutaraldehyde	111-30-8	20
Glycerol (1,2,3-Propanetriol, Glycerine)	56-81-5	16
Glycol ethers	*	9
Glycol ethers	9004-77-7	4
Glyoxal	107-22-2	3
Glyoxylic acid	298-12-4	1
Guar gum	9000-30-0	41
Guar gum derivative	*	12
Haloalkyl heteropolycycle salt	*	6
Heavy aromatic distillate	68132-00-3	1
Heavy aromatic petroleum naphtha	64742-94-5	45
Heavy catalytic reformed petroleum naphtha	64741-68-0	10
Hematite	*	5
Hemicellulase	9025-56-3	2
Hexahydro-1,3,5-tris(2-hydroxyethyl)-s-triazine (Triazine)	4719-04-4	4
Hexamethylenetetramine	100-97-0	37
Hexanediamine	124-09-4	1
Hexanes	*	1
Hexylene glycol	107-41-5	5
Hydrated aluminum silicate	1332-58-7	4
Hydrocarbon mixtures	8002-05-9	1
Hydrocarbons	*	3
Hydrodesulfurized kerosine (petroleum)	64742-81-0	3
Hydrodesulfurized light catalytic cracked distillate (petroleum)	68333-25-5	1

Appendix A. (Continued).

Chemical Component	Chemical Abstract Service Number	No. of Products Containing Chemical
Hydrodesulfurized middle distillate (petroleum)	64742-80-9	1
Hydrogen chloride (Hydrochloric acid)	7647-01-0	42
Hydrogen fluoride (Hydrofluoric acid)	7664-39-3	2
Hydrogen peroxide	7722-84-1	4
Hydrogen sulfide	7783-06-4	1
Hydrotreated and hydrocracked base oil	*	2
Hydrotreated heavy naphthenic distillate	64742-52-5	3
Hydrotreated heavy paraffinic petroleum distillates	64742-54-7	1
Hydrotreated heavy petroleum naphtha	64742-48-9	7
Hydrotreated light petroleum distillates	64742-47-8	89
Hydrotreated middle petroleum distillates	64742-46-7	3
Hydroxyacetic acid (Glycolic acid)	79-14-1	6
Hydroxyethylcellulose	9004-62-0	1
Hydroxyethylethylenediaminetriacetic acid, trisodium salt	139-89-9	1
Hydroxylamine hydrochloride	5470-11-1	1
Hydroxypropyl guar gum	39421-75-5	2
Hydroxysultaine	*	1
Inner salt of alkyl amines	*	2
Inorganic borate	*	3
Inorganic particulate	*	1
Inorganic salt	*	1
Inorganic salt	533-96-0	1
Inorganic salt	7446-70-0	1
Instant coffee purchased off the shelf	*	1
Inulin, carboxymethyl ether, sodium salt	430439-54-6	1
Iron oxide	1332-37-2	2
Iron oxide (Ferric oxide)	1309-37-1	18
Iso amyl alcohol	123-51-3	1
Iso-alkanes/n-alkanes	*	10
Isobutanol (Isobutyl alcohol)	78-83-1	4

Chemical Component	Chemical Abstract Service Number	No. of Products Containing Chemical
Isomeric aromatic ammonium salt	*	1
Isooctanol	26952-21-6	1
Isooctyl alcohol	68526-88-0	1
Isooctyl alcohol bottoms	68526-88-5	1
Isopropanol (Isopropyl alcohol, Propan-2-ol)	67-63-0	274
Isopropylamine	75-31-0	1
Isotridecanol, ethoxylated	9043-30-5	1
Kerosene	8008-20-6	13
Lactic acid	10326-41-7	1
Lactic acid	50-21-5	1
L-Dilactide	4511-42-6	1
Lead	7439-92-1	1
Light aromatic solvent naphtha	64742-95-6	11
Light catalytic cracked petroleum distillates	64741-59-9	1
Light naphtha distillate, hydrotreated	64742-53-6	1
Low toxicity base oils	*	1
Maghemite	*	2
Magnesium carbonate	546-93-0	1
Magnesium chloride	7786-30-3	4
Magnesium hydroxide	1309-42-8	4
Magnesium iron silicate	1317-71-1	3
Magnesium nitrate	10377-60-3	5
Magnesium oxide	1309-48-4	18
Magnesium peroxide	1335-26-8	2
Magnesium peroxide	14452-57-4	4
Magnesium phosphide	12057-74-8	1
Magnesium silicate	1343-88-0	3
Magnesium silicate hydrate (talc)	14807-96-6	2
Magnetite	*	3
Medium aliphatic solvent petroleum naphtha	64742-88-7	10
Metal salt	*	2
Metal salt solution	*	1
Methanol (Methyl alcohol)	67-56-1	342

Appendix A. (Continued).

Chemical Component	Chemical Abstract Service Number	No. of Products Containing Chemical
Methyl isobutyl carbinol (Methyl amyl alcohol)	108-11-2	3
Methyl salicylate	119-36-8	6
Methyl vinyl ketone	78-94-4	2
Methylcyclohexane	108-87-2	1
Mica	12001-26-2	3
Microcrystalline silica	1317-95-9	1
Mineral	*	1
Mineral Filler	*	1
Mineral spirits (stoddard solvent)	8052-41-3	2
Mixed titanium ortho ester complexes	*	1
Modified alkane	*	1
Modified cycloaliphatic amine adduct	*	3
Modified lignosulfonate	*	1
Monoethanolamine (Ethanolamine)	141-43-5	17
Monoethanolamine borate	26038-87-9	1
Morpholine	110-91-8	2
Mullite	1302-93-8	55
n,n-dibutylthiourea	109-46-6	1
N,N-dimethyl-1-octadecanamine-HCl	*	1
N,N-dimethyloctadecylamine	124-28-7	3
N,N-dimethyloctadecylamine hydrochloride	1613-17-8	2
n,n'-Methylenebisacrylamide	110-26-9	1
n-alkyl dimethyl benzyl ammonium chloride	139-08-2	1
Naphthalene	91-20-3	44
Naphthalene derivatives	*	1
Naphthalenesulphonic acid, bis (1-methylethyl)-methyl derivatives	99811-86-6	1
Natural asphalt	12002-43-6	1
n-cocoamidopropyl-n,n-dimethyl-n-2-hydroxypropylsulfobetaine	68139-30-0	1
n-dodecyl-2-pyrrolidone	2687-96-9	1
N-heptane	142-82-5	1

Chemicals Used in Hydraulic Fracturing

Chemical Component	Chemical Abstract Service Number	No. of Products Containing Chemical
Nickel sulfate hexahydrate	10101-97-0	2
Nitrilotriacetamide	4862-18-4	4
Nitrilotriacetic acid	139-13-9	6
Nitrilotriacetonitrile	7327-60-8	3
Nitrogen	7727-37-9	9
n-Methylpyrrolidone	872-50-4	1
Nonane, all isomers	*	1
Non-hazardous salt	*	1
Nonionic surfactant	*	1
Nonyl phenol ethoxylate	*	2
Nonyl phenol ethoxylate	9016-45-6	2
Nonyl phenol ethoxylate	9018-45-9	1
Nonylphenol	25154-52-3	1
Nonylphenol, ethoxylated and sulfated	9081-17-8	1
N-propyl zirconate	*	1
N-tallowalkyltrimethylenediamines	*	1
Nuisance particulates	*	2
Nylon fibers	25038-54-4	2
Octanol	111-87-5	2
Octyltrimethylammonium bromide	57-09-0	1
Olefinic sulfonate	*	1
Olefins	*	1
Organic acid salt	*	3
Organic acids	*	1
Organic phosphonate	*	1
Organic phosphonate salts	*	1
Organic phosphonic acid salts	*	6
Organic salt	*	1
Organic sulfur compound	*	2
Organic titanate	*	2
Organiophilic clay	*	2
Organo-metallic ammonium complex	*	1
Other inorganic compounds	*	1
Oxirane, methyl-, polymer with oxirane, mono-C10-16-alkyl ethers, phosphates	68649-29-6	1
Oxyalkylated alcohol	*	6

Appendix A. (Continued).

Chemical Component	Chemical Abstract Service Number	No. of Products Containing Chemical
Oxyalkylated alcohols	228414-35-5	1
Oxyalkylated alkyl alcohol	*	1
Oxyalkylated alkylphenol	*	1
Oxyalkylated fatty acid	*	2
Oxyalkylated phenol	*	1
Oxyalkylated polyamine	*	1
Oxylated alcohol	*	1
Paraffin wax	8002-74-2	1
Paraffinic naphthenic solvent	*	1
Paraffinic solvent	*	5
Paraffins	*	1
Perlite	93763-70-3	1
Petroleum distillates	*	26
Petroleum distillates	64742-65-0	1
Petroleum distillates	64742-97-5	1
Petroleum distillates	68477-31-6	3
Petroleum gas oils	*	1
Petroleum gas oils	64741-43-1	1
Phenol	108-95-2	5
Phenol-formaldehyde resin	9003-35-4	32
Phosphate ester	*	6
Phosphate esters of alkyl phenyl ethoxylate	68412-53-3	1
Phosphine	*	1
Phosphonic acid	*	1
Phosphonic acid	129828-36-0	1
Phosphonic acid	13598-36-2	3
Phosphonic acid (dimethlamino(methylene))	29712-30-9	1
Phosphonic acid, [nitrilotris(methylene)]tris-,	2235-43-0	1
Phosphoric acid	7664-38-2	7
Phosphoric acid ammonium salt	*	1
Phosphoric acid, mixed decyl, octyl and ethyl esters	68412-60-2	3
Phosphorous acid	10294-56-1	1
Phthalic anhydride	85-44-9	2

Chemicals Used in Hydraulic Fracturing

Chemical Component	Chemical Abstract Service Number	No. of Products Containing Chemical
Pine oil	8002-09-3	5
Plasticizer	*	1
Poly(oxy-1,2-ethanediyl)	24938-91-8	1
Poly(oxy-1,2-ethanediyl), alpha-(4-nonylphenyl)-omega-hydroxy-, branched (Nonylphenol ethoxylate)	127087-87-0	3
Poly(oxy-1,2-ethanediyl), alpha-hydro-omega-hydroxy	65545-80-4	1
Poly(oxy-1,2-ethanediyl), alpha-sulfo-omega-(hexyloxy)-, ammonium salt	63428-86-4	3
Poly(oxy-1,2-ethanediyl),a-(nonylphenyl)-w-hydroxy-, phosphate	51811-79-1	1
Poly-(oxy-1,2-ethanediyl)-alpha-undecyl-omega-hydroxy	34398-01-1	6
Poly(sodium-p-styrenesulfonate)	25704-18-1	1
Poly(vinyl alcohol)	25213-24-5	2
Polyacrylamides	9003-05-8	2
Polyacrylamides	*	1
Polyacrylate	*	1
Polyamine	*	2
Polyanionic cellulose	*	2
Polyepichlorohydrin, trimethylamine quaternized	51838-31-4	1
Polyetheramine	9046-10-0	3
Polyether-modified trisiloxane	27306-78-1	1
Polyethylene glycol	25322-68-3	20
Polyethylene glycol ester with tall oil fatty acid	9005-02-1	1
Polyethylene polyammonium salt	68603-67-8	2
Polyethylene-polypropylene glycol	9003-11-6	5
Polylactide resin	*	3
Polyoxyalkylenes	*	1
Polyoxyethylene castor oil	61791-12-6	1
Polyphosphoric acid, esters with triethanolamine, sodium salts	68131-72-6	1
Polypropylene glycol	25322-69-4	1

Appendix A. (Continued).

Chemical Component	Chemical Abstract Service Number	No. of Products Containing Chemical
Polysaccharide	*	20
Polyvinyl alcohol	*	1
Polyvinyl alcohol	9002-89-5	2
Polyvinyl alcohol/polyvinylacetate copolymer	*	1
Potassium acetate	127-08-2	1
Potassium carbonate	584-08-7	12
Potassium chloride	7447-40-7	29
Potassium formate	590-29-4	3
Potassium hydroxide	1310-58-3	25
Potassium iodide	7681-11-0	6
Potassium metaborate	13709-94-9	3
Potassium metaborate	16481-66-6	3
Potassium oxide	12136-45-7	1
Potassium pentaborate	*	1
Potassium persulfate	7727-21-1	9
Propanol (Propyl alcohol)	71-23-8	18
Propanol, [2(2-methoxy-methylethoxy) methylethoxyl]	20324-33-8	1
Propargyl alcohol (2-propyn-1-ol)	107-19-7	46
Propylene carbonate (1,3-dioxolan-2-one, methyl-)	108-32-7	2
Propylene glycol (1,2-propanediol)	57-55-6	18
Propylene oxide	75-56-9	1
Propylene pentamer	15220-87-8	1
p-Xylene	106-42-3	1
Pyridinium, 1-(phenylmethyl)-, ethyl methyl derivatives, chlorides	68909-18-2	9
Pyrogenic silica	112945-52-5	3
Quaternary amine compounds	*	3
Quaternary amine compounds	61789-18-2	1
Quaternary ammonium compounds	*	9
Quaternary ammonium compounds	19277-88-4	1
Quaternary ammonium compounds	68989-00-4	1

Chemicals Used in Hydraulic Fracturing

Chemical Component	Chemical Abstract Service Number	No. of Products Containing Chemical
Quaternary ammonium compounds	8030-78-2	1
Quaternary ammonium compounds, dicoco alkyldimethyl, chlorides	61789-77-3	2
Quaternary ammonium salts	*	2
Quaternary compound	*	1
Quaternary salt	*	2
Quaternized alkyl nitrogenated compound	68391-11-7	2
Rafinnates (petroleum), sorption process	64741-85-1	2
Residues (petroleum), catalytic reformer fractionator	64741-67-9	10
Resin	8050-09-7	2
Rutile	1317-80-2	2
Salt of phosphate ester	*	3
Salt of phosphono-methylated diamine	*	1
Salts of oxyalkylated fatty amines	68551-33-7	1
Secondary alcohol	*	7
Silica (Silicon dioxide)	7631-86-9	47
Silica, amorphous	*	3
Silica, amorphous precipitated	67762-90-7	1
Silicon carboxylate	681-84-5	1
Silicon dioxide (Fused silica)	60676-86-0	7
Silicone emulsion	*	1
Sodium (C14-16) olefin sulfonate	68439-57-6	4
Sodium 2-ethylhexyl sulfate	126-92-1	1
Sodium acetate	127-09-3	6
Sodium acid pyrophosphate	7758-16-9	5
Sodium alkyl diphenyl oxide sulfonate	28519-02-0	1
Sodium aluminate	1302-42-7	1
Sodium aluminum phosphate	7785-88-8	1
Sodium bicarbonate (Sodium hydrogen carbonate)	144-55-8	10
Sodium bisulfite	7631-90-5	6
Sodium bromate	7789-38-0	10
Sodium bromide	7647-15-6	1
Sodium carbonate	497-19-8	14

Appendix A. (Continued).

Chemical Component	Chemical Abstract Service Number	No. of Products Containing Chemical
Sodium chlorate	7775-09-9	1
Sodium chloride	7647-14-5	48
Sodium chlorite	7758-19-2	8
Sodium cocaminopropionate	68608-68-4	2
Sodium diacetate	126-96-5	2
Sodium erythorbate	6381-77-7	4
Sodium glycolate	2836-32-0	2
Sodium hydroxide (Caustic soda)	1310-73-2	80
Sodium hypochlorite	7681-52-9	14
Sodium lauryl-ether sulfate	68891-38-3	3
Sodium metabisulfite	7681-57-4	1
Sodium metaborate	7775-19-1	2
Sodium metaborate tetrahydrate	35585-58-1	6
Sodium metasilicate, anhydrous	6834-92-0	2
Sodium nitrite	7632-00-0	1
Sodium oxide (Na2O)	1313-59-3	1
Sodium perborate	1113-47-9	1
Sodium perborate	7632-04-4	1
Sodium perborate tetrahydrate	10486-00-7	4
Sodium persulfate	7775-27-1	6
Sodium phosphate	*	2
Sodium polyphosphate	68915-31-1	1
Sodium salicylate	54-21-7	1
Sodium silicate	1344-09-8	2
Sodium sulfate	7757-82-6	7
Sodium tetraborate	1330-43-4	7
Sodium tetraborate decahydrate	1303-96-4	10
Sodium thiosulfate	7772-98-7	10
Sodium thiosulfate pentahydrate	10102-17-7	3
Sodium trichloroacetate	650-51-1	1
Sodium tripolyphosphate	7758-29-4	2
Sodium xylene sulfonate	1300-72-7	3
Sodium zirconium lactate	174206-15-6	1

Chemicals Used in Hydraulic Fracturing

Chemical Component	Chemical Abstract Service Number	No. of Products Containing Chemical
Solvent refined heavy naphthenic petroleum	64741-96-4	1
Sorbitan monooleate	1338-43-8	1
Stabilized aqueous chlorine dioxide	10049-04-4	1
Stannous chloride	7772-99-8	1
Stannous chloride dihydrate	10025-69-1	6
Starch	9005-25-8	5
Steam cracked distillate, cyclodiene dimer, dicyclopentadiene polymer	68131-87-3	1
Steam-cracked petroleum distillates	64742-91-2	6
Straight run middle petroleum distillates	64741-44-2	5
Substituted alcohol	*	2
Substituted alkene	*	1
Substituted alkylamine	*	2
Sucrose	57-50-1	1
Sulfamic acid	5329-14-6	6
Sulfate	*	1
Sulfonate acids	*	1
Sulfonate surfactants	*	1
Sulfonic acid salts	*	1
Sulfonic acids, petroleum	61789-85-3	1
Sulfur compound	*	1
Sulfuric acid	7664-93-9	9
Sulfuric acid, monodecyl ester, sodium salt	142-87-0	2
Sulfuric acid, monooctyl ester, sodium salt	142-31-4	2
Surfactants	*	13
Sweetened middle distillate	64741-86-2	1
Synthetic organic polymer	9051-89-2	2
Tall oil (Fatty acids)	61790-12-3	4
Tall oil, compound with diethanolamine	68092-28-4	1
Tallow soap	*	2
Tar bases, quinoline derivatives, benzyl chloride-quaternized	72480-70-7	5
Tergitol	68439-51-0	1
Terpene hydrocarbon byproducts	68956-56-9	3
Terpenes	*	1
Terpenes and terpenoids, sweet orange-oil	68647-72-3	2

Appendix A. (Continued).

Chemical Component	Chemical Abstract Service Number	No. of Products Containing Chemical
Terpineol	8000-41-7	1
Tert-butyl hydroperoxide	75-91-2	6
Tetra-calcium-alumino-ferrite	12068-35-8	1
Tetraethylene glycol	112-60-7	1
Tetraethylenepentamine	112-57-2	2
Tetrahydro-3,5-dimethyl-2H-1,3,5-thiadiazine-2-thione (Dazomet)	533-74-4	13
Tetrakis (hydroxymethyl) phosphonium sulfate	55566-30-8	12
Tetramethyl ammonium chloride	75-57-0	14
Tetrasodium 1-hydroxyethylidene-1,1-diphosphonic acid	3794-83-0	1
Tetrasodium ethylenediaminetetraacetate	64-02-8	10
Thiocyanate sodium	540-72-7	1
Thioglycolic acid	68-11-1	6
Thiourea	62-56-6	9
Thiourea polymer	68527-49-1	3
Titanium complex	*	1
Titanium oxide	13463-67-7	19
Titanium, isopropoxy (triethanolaminate)	74665-17-1	2
Toluene	108-88-3	29
Treated ammonium chloride (with anti-caking agent a or b)	12125-02-9	1
Tributyl tetradecyl phosphonium chloride	81741-28-8	5
Tri-calcium silicate	12168-85-3	1
Tridecyl alcohol	112-70-9	1
Triethanolamine (2,2,2-nitrilotriethanol)	102-71-6	21
Triethanolamine polyphosphate ester	68131-71-5	3
Triethanolamine titanate	36673-16-2	1
Triethanolamine zirconate	101033-44-7	6
Triethanolamine zirconium chelate	*	1
Triethyl citrate	77-93-0	1
Triethyl phosphate	78-40-0	1

Chemical Component	Chemical Abstract Service Number	No. of Products Containing Chemical
Triethylene glycol	112-27-6	3
Triisopropanolamine	122-20-3	5
Trimethylammonium chloride	593-81-7	1
Trimethylbenzene	25551-13-7	5
Trimethyloctadecylammonium (1-octadecanaminium, N,N,N-trimethyl-, chloride)	112-03-8	6
Tris(hydroxymethyl)aminomethane	77-86-1	1
Trisodium ethylenediaminetetraacetate	150-38-9	1
Trisodium ethylenediaminetriacetate	19019-43-3	1
Trisodium nitrilotriacetate	18662-53-8	8
Trisodium nitrilotriacetate (Nitrilotriacetic acid, trisodium salt monohydrate)	5064-31-3	9
Trisodium ortho phosphate	7601-54-9	1
Trisodium phosphate dodecahydrate	10101-89-0	1
Ulexite	1319-33-1	1
Urea	57-13-6	3
Wall material	*	1
Walnut hulls	*	2
White mineral oil	8042-47-5	8
Xanthan gum	11138-66-2	6
Xylene	1330-20-7	44
Zinc chloride	7646-85-7	1
Zinc oxide	1314-13-2	2
Zirconium complex	*	10
Zirconium dichloride oxide	7699-43-6	1
Zirconium oxide sulfate	62010-10-0	2
Zirconium sodium hydroxy lactate complex	113184-20-6	2

* Components marked with an asterisk appeared on at least one MSDS without an identifying CAS number. The MSDSs in these cases marked the CAS as proprietary, noted that the CAS was not available, or left the CAS field blank. Components marked with an asterisk may be duplicative of other components on this list, but Committee staff have no way of identifying such duplicates without the identifying CAS number.

End Notes

[1] Energy Information Administration (EIA), *Natural Gas Monthly (Mar. 2011)*, Table 1, U.S. Natural Gas Monthly Supply and Disposition Balance (online at www.eia.gov/dnav/ng/hist/n9070us1A.htm) (accessed Mar. 30, 2011).

[2] EIA, *Annual Energy Outlook 2011 Early Release* (Dec. 16, 2010); EIA, *What is shale gas and why is it important?* (online at www.eia.doe.gov/energy (accessed Mar. 30, 2011).

[3] U.S. Environmental Protection Agency, *Evaluation of Impacts to Underground Sources of Drinking Water by Hydraulic Fracturing of Coalbed Methane Reservoirs* (June 2004) (EPA 816-R-04-003) at 4-1 and 4-2.

[4] For instance, Pennsylvania's Department of Environmental Protection has cited Cabot Oil & Gas Corporation for contamination of drinking water wells with seepage caused by weak casing or improper cementing of a natural gas well. *See Officials in Three States Pin Water Woes on Gas Drilling*, ProPublica (Apr. 26, 2009) (online at www.propublica.org/article/officials (accessed Mar. 24, 2011).

[5] John A. Veil, Argonne National Laboratory, *Water Management Technologies Used by Marcellus Shale Gas Producers*, prepared for the Department of Energy (July 2010), at 13 (hereinafter "*Water Management Technologies*").

[6] 42 U.S.C. § 300h(d). Many dubbed this provision the "Halliburton loophole" because of Halliburton's ties to then-Vice President Cheney and its role as one of the largest providers of hydraulic fracturing services. *See The Halliburton Loophole*, New York Times (Nov. 9. 2009).

[7] *See* EPA, *Draft Hydraulic Fracturing Study Plan* (Feb. 7, 2011), at 37; *Regulation Lax as Gas Wells' Tainted Water Hits Rivers*, New York Times (Feb. 26, 2011).

[8] *Water Management Technologies*, at 13.

[9] *Regulation Lax as Gas Wells' Tainted Water Hits Rivers*, New York Times (Feb. 26, 2011).

[10] Wyoming, for example, recently enacted relatively strong disclosure regulations, requiring disclosure on a well-by-well basis and "for each stage of the well stimulation program," "the chemical additives, compounds and concentrations or rates proposed to be mixed and injected." *See* WCWR 055-000-003 Sec. 45. Similar regulations became effective in Arkansas this year. *See* Arkansas Oil and Gas Commission Rule B-19. In Wyoming, much of this information is, after an initial period of review, available to the public. *See* WCWR 055 000-003 Sec. 21. Other states, however, do not insist on such robust disclosure. For instance, West Virginia has no disclosure requirements for hydraulic fracturing and expressly exempts fluids used during hydraulic fracturing from the disclosure requirements applicable to underground injection of fluids for purposes of waste storage. *See* W. Va. Code St. R. § 34-5-7.

[11] *See Ground Water Protection Council Calls for Disclosure of Chemicals Used in Shale Gas Exploration*, Ground Water Protection Council (Oct. 5, 2010) (online at www.wqpmag.com/Ground-Water-Protection-Council-Calls-for-Disclosure-of-Chemicals-inShale-Gas-Exploration-newsPiece21700) (accessed Mar. 24, 2011).

[12] The Committee sent letters to Basic Energy Services, BJ Services, Calfrac Well Services, Complete Production Services, Frac Tech Services, Halliburton, Key Energy Services, RPC, Sanjel Corporation, Schlumberger, Superior Well Services, Trican Well Service, Universal Well Services, and Weatherford.

[13] BJ Services, Halliburton, and Schlumberger already had provided the Oversight Committee with data for 2005 through 2007. For BJ Services, the 2005-2007 data is limited to natural gas wells. For Schlumberger, the 2005-2007 data is limited to coalbed methane wells.

[14] 29 CFR 1910.1200(g)(2)(i)(C)(1).

[15] 29 CFR 1910.1200.

[16] Each hydraulic fracturing "product" is a mixture of chemicals or other components designed to achieve a certain performance goal, such as increasing the viscosity of water. Some oil and gas service companies create their own products; most purchase these products from chemical vendors. The service companies then mix these products together at the well site to formulate the hydraulic fracturing fluids that they pump underground.

[17] EPA, *Toxicological Review of Ethylene Glycol Monobutyl Ether* (Mar. 2010) at 4.

[18] EPA, *Fact Sheet: January 2010 Sampling Results and Site Update, Pavillion, Wyoming Groundwater Investigation* (Aug. 2010) (online at www.epa.gov/region8/ superfund/ wy/ pavillion/PavillionWyomingFactSheet.pdf) (accessed Mar. 1, 2011).

[19] According to EPA, diesel contains benzene, toluene, ethylbenzene, and xylenes. See EPA, Evaluation of Impacts to Underground Sources of Drinking Water by Hydraulic Fracturing of Coalbed Methane Reservoirs (June 2004) (EPA 816-R-04-003) at 4-11.

[20] For purposes of this report, a chemical is considered a "carcinogen" if it is on one of two lists: (1) substances identified by the National Toxicology Program as "known to be human carcinogens" or as "reasonably anticipated to be human carcinogens"; and (2) substances identified by the International Agency for Research on Cancer, part of the World Health Organization, as "carcinogenic" or "probably carcinogenic" to humans. See U.S. Department of Health and Human Services, Public Health Service, National Toxicology Program, *Report on Carcinogens, Eleventh Edition* (Jan. 31, 2005) and World Health Organization, International Agency for Research on Cancer, *Agents Classified by the IARC Monographs* (online at http://monographs.iarc.fr/ENG/Classification/index.php) (accessed Feb. 28, 2011).

[21] U.S. Department of Health and Human Services, Agency for Toxic Substances and Disease Registry, *Public Health Statement for Benzene* (Aug. 2007).

[22] EPA, *Basic Information about Toluene in Drinking Water, Basic Information about Ethylbenzene in Drinking Water*, and *Basic Information about Xylenes in Drinking Water* (online at http://water (accessed Oct. 14, 2010).

[23] Letter from Reps. Henry A. Waxman, Edward J. Markey, and Diana DeGette to the Honorable Lisa Jackson, Administrator, U.S. Environmental Protection Agency (Jan. 31, 2011).

[24] EPA, *Evaluation of Impacts to Underground Sources of Drinking Water by Hydraulic Fracturing of Coalbed Methane Reservoirs* (June 2004) (EPA 816-R-04-003) at 4-11.

[25] *Id.*

[26] EPA, *Contaminant Candidate List 3* (online at http://water.epa.gov/scitech/drinkingwater/ dws/ccl/ ccl3.cfm) (accessed Mar. 31, 2011).

[27] Clean Air Act Section 112(b), 42 U.S.C. § 7412.

[28] HHS, Agency for Toxic Substances and Disease Registry, *Medical Management Guidelines for Hydrogen Fluoride* (online at www.atsdr.cdc.gov/mhmi/mmg11.pdf) (accessed Mar. 24, 2011).

[29] EPA, *Basic Information about Lead* (online at www.epa.gov/lead (accessed Mar. 30, 2011).

[30] This is likely a conservative estimate. We included only those products for which the MSDS says "proprietary" or "trade secret" instead of listing a component by name or providing the CAS number. If the MSDS listed a component's CAS as N.A. or left it blank, we did not count that as a trade secret claim, unless the company specified as such in follow-up correspondence.

[31] Letter from Reginald J. Brown to Henry A. Waxman, Chairman, Committee on Energy and Commerce, and Edward J. Markey, Chairman, Subcommittee on Energy and Environment (Apr. 16, 2010).
[32] Letter from Philip Perry to Henry A. Waxman, Chairman, Committee Energy and Commerce, and Edward J. Markey, Chairman, Subcommittee on Energy and Environment (Aug. 6, 2010).
[33] E-mail from Peter Spivack to Committee Staff (Aug. 5, 2010).
[34] E-mail from Lee Blalack to Committee Staff (July 29, 2010).
[35] To compile this list of chemicals, Committee staff reviewed each Material Safety Data Sheet provided to the Committee for hydraulic fracturing products used between 2005 and 2009. Committee staff transcribed the names and CAS numbers as written in the MSDSs; as such, any inaccuracies on this list reflect inaccuracies on the MSDSs themselves.

Chapter 3

STATEMENT OF BARBARA BOXER, BEFORE THE SUBCOMMITTEE ON WATER AND WILDLIFE, HEARING ON "NATURAL GAS DRILLING: PUBLIC HEALTH AND ENVIRONMENTAL IMPACTS"[*]

Today we will examine the public health and environmental impacts of natural gas drilling. This hearing comes at an important time in the history of natural gas development in the U.S.

Recent advancements in horizontal drilling and hydraulic fracturing have led to a significant expansion in proven U.S. natural gas reserves. Natural gas resources are now recoverable that were considered uneconomical even a few years ago.

The discovery of new resources creates an opportunity for increased production of a cleaner, domestically-produced fuel. While this could have benefits for our nation's economy and energy independence, it is critical to ensure that exploration for natural gas is done safely and responsibly.

One of the key reasons for the increase in natural gas reserves is the discovery of the Marcellus Shale in the Appalachian region of the United States, which underlays portions of Virginia, West Virginia, Ohio, Maryland, Pennsylvania, and New York. With drilling in this part of the country likely to

---

[*] This is an edited, reformatted and augmented version of Statement given by Barbara Boxer, before the Subcommittee on Water and Wildlife, Hearing on "Natural Gas Drilling: Public Health and Environmental Impacts" on Tuesday, April 12, 2011.

increase exponentially in coming years, it is critical that we ensure efforts to extract natural gas do not threaten the air we breathe and the water we drink.

I want to thank Senator Cardin, whose state is directly impacted by the Marcellus shale discovery, for his leadership on this important issue and for helping to spearhead our Committee's oversight efforts on public health and environmental impacts of natural gas exploration. This oversight is vital to make sure drilling is done responsibly and not in a way that would subject our children and families to harmful pollution.

There are questions that need to be answered about the safety of natural gas exploration and hydraulic fracturing. A recent series of investigative reports in the New York Times highlighted the potential risks of natural gas drilling and inconsistent efforts to regulate this booming industry. Some of the findings of the New York Times investigation raise significant issues.

For example, the Times reported that hydraulic fracturing process wastewater is often contaminated with pollutants, including toxic metals, highly corrosive salts, carcinogens such as benzene, and radioactive elements. A large amount of this wastewater is disposed in municipal sewage treatment plants that may not be equipped to remove the contaminants.

These plants can discharge harmful levels of radiation and toxic substances into local waterways, and the solid waste produced may also contain an array of toxins. Without proper oversight, the disposal of drilling wastewater poses threats to both aquatic life and human health, especially when public drinking water systems rely on waterways where waste is being discharged.

Concerns have also been raised that chemicals contained in the fluids used in the hydraulic fracturing process can contaminate groundwater sources. However, federal and state regulators and concerned citizens have not had all the information they need to determine whether drilling is causing groundwater contamination. Historically, some companies have limited access to information on the chemicals used in their drilling fluids.

The Federal government does not currently require drilling operators to fully disclose the chemicals they are injecting into the ground. Some states, such as Wyoming, now require disclosure of chemicals used in the fracking process. The industry has also recently launched a voluntary disclosure effort with the Ground Water Protection Council and others. These are encouraging developments, but we still have a long way to go before full disclosure is a consistent, industry-wide practice.

I have highlighted only a few of the health and environmental issues that have been associated with natural gas drilling. Additional issues include air pollution and impacts on water supplies due to the millions of gallons of water that are needed at each natural gas well.

Given the array of potential impacts and the need for more study, the state of New York is taking a time-out on hydraulic fracturing, choosing to fully study the issues first before allowing widespread drilling.

The U.S. EPA has also been directed by Congress to study the impacts of hydraulic fracturing on water supplies. I expect that the agency will use an independent, comprehensive and scientific process to provide Congress and the public with accurate and unbiased information that will help us ensure public health is protected.

There is much still to be learned about the impacts of natural gas drilling. With the rapid expansion of exploration for this energy source and the potential for it to help meet a significant portion of the nation's energy needs, we have much to gain from a full review of this issue. We must move forward in a way that ensures safe and responsible drilling that is protective of our air and water.

This hearing is an important step in the EPW committee's oversight on this issue. I look forward to the testimony of the witnesses here today.

Chapter 4

STATEMENT OF BENJAMIN L. CARDIN, BEFORE THE SUBCOMMITTEE ON WATER AND WILDLIFE, HEARING ON "NATURAL GAS DRILLING: PUBLIC HEALTH AND ENVIRONMENTAL IMPACTS"[*]

The United States has as much natural gas as Saudi Arabia has oil. According to Penn State, the Marcellus Shale, which runs from central New York State to West Virginia, may be the second largest natural gas field in the world.

We have enormous reserves that can help America meet its energy needs and do so in a way that produces far less pollution than coal, helps the United States on its path to energy independence, and improves national security.

High volume, horizontal hydraulic fracturing, or "fracking," is now being used to extract natural gas from shale formations in thousands of new wells. In Pennsylvania more than 2,700 Marcellus wells were drilled from 2006 to March 10th of this year.

A study last year estimated that Marcellus drilling would create or support more than 100,000 jobs in 2011, plus billions of dollars in economic value for the state.

The natural gas industry is booming, but that may soon end. Against the backdrop of natural gas's promise,

---

[*] This is an edited, reformatted and augmented version of testimony given by Benjamin L. Cardin, before the Subcommittee on Water and Wildlife, Hearing on "Natural Gas Drilling: Public Health and Environmental Impacts" on Tuesday, April 12, 2011.

- New York and Maryland have imposed moratoria on fracking operations.
- New Jersey is considering a ban on the practice.
- The City of Pittsburgh has enacted a ban on fracking operations within the city limits.
- Tiny Mountain Lake Park, Maryland adopted an ordinance making the drilling of natural gas illegal within the town limits.

What's going on? In the face of its extraordinary promise, why is natural gas faltering?

The answer is simple. The industry has failed to meet minimally acceptable performance levels for protecting human health and the environment. That is both an industry failure and a failure of the regulatory agencies.

I am a strong supporter of domestic natural gas production. But my support only comes when human health and the environment are protected.

POTENTIAL HUMAN HEALTH AND ENVIRONMENTAL IMPACTS

The natural gas industry argues that there has never been a documented case of drinking water contamination from fracking.

Viewed in isolation, fracking occurs far below drinking water aquifers. But fracking doesn't mysteriously just happen. It involves drilling, wells, water, compressors, and all the associated equipment that goes into a modern well-drilling operation.

The record is replete with cases of contamination from improper cement jobs, cracked drill casings, drill pad spills, and seismic disturbances releasing natural gas in higher geological formations. For example, Pennsylvania DEP brought an enforcement action against Cabot Oil for poor cementing after the drinking water wells of 19 families in Dimock, PA were polluted.

In June, 2010 the Pennsylvania Land Trust Association identified a total of 1614 violations accrued by 45 Pennsylvania Marcellus Shale drillers, dating to January 2008, including:

- 91 violations of Pennsylvania's Clean Streams Law
- 162 cases of Improper Construction of Waste Water Impoundments
- 50 cases of Improper Well-Casing Construction, and
- 4 cases of inadequate Blowout Prevention. Last June a well blowout in Clearfield County shot 35,000 gallons of gas and water 75 ft into the air over a 16 hour period.

Pro Publica, the investigative news site, has found over 1,000 reports of water contamination near drilling sites.

Up to 5 million gallons of water combined with thousands of gallons of special chemicals can be used in a single fracking operation, much of this returning to the surface.

This flowback fluid contains not only the chemicals used in the fracking process, but can also include salts, metals, and naturally occurring radioactive substances.

Municipal wastewater treatment plants are not equipped to handle these contaminants. In Pennsylvania, more than a dozen publically owned treatment works facilities accepting natural gas wastewater have failed to treat it in accordance with permit requirements. Radioactive isotopes, heavy salts, and other chemicals were discharged into surface waters – all in apparent violation of the Clean Water Act.

Even specialized facilities can have problems.

Studies of the effluent from a commercial facility in Pennsylvania that collects water only from gas operations show half a dozen pollutants in excess of their approved limits as determined by an agency at the Centers for Disease Control.

The contaminants include:

- Chloride, which has been found at levels that are hundreds of times higher than accepted health levels.
- Benzene, a known carcinogen, has been detected at levels thousands of times above health limits.
- Bromide reacts with other chemicals to form potentially carcinogenic compounds. It is present at levels tens of thousands of times higher than the levels of concern.

Under the Safe Drinking Water Act, the EPA would typically be allowed to regulate all underground injections of fluids, including the chemicals used in fracking, cement jobs, casings, and disposal of flowback water. However, a loophole exempts fracking from regulation, except where diesel fuel is used.

Even with this huge loophole, federal violations have occurred. Since 2005, companies have injected over 32 million gallons of diesel fuel or fracking fluids containing diesel fuel in wells in 19 states. None of these operations obtained a permit under the Safe Drinking Water Act, meaning that all are in violation. To date, EPA has failed to take any enforcement action.

Similar restrictions on the Clean Water Act allow drilling companies to operate outside its scope.

State regulators, facing their own massive budget cuts, have tried to fill the void. In the Marcellus Shale area, Pennsylvania's response has been characterized as playing continual regulatory catch-up, as regulations have routinely failed to address issues.

As today's hearing will make clear, the exemptions from the Safe Drinking Water Act and Clean Water Act aren't working.

We need to put the environmental cop back on the beat, take aggressive action against the bad actors in the industry and earn back the public's confidence.

The promise of natural gas will be a promise unfulfilled if the human health and environmental impacts are not properly safe-guarded. It's long past time that they were.

Chapter 5

STATEMENT OF JAMES M. INHOFE, BEFORE THE SUBCOMMITTEE ON WATER AND WILDLIFE, HEARING ON "NATURAL GAS DRILLING: PUBLIC HEALTH AND ENVIRONMENTAL IMPACTS"[*]

On March 17, 1949, more than 60 years ago, the first hydraulic fracturing job was performed on a well 12 miles east of Duncan, in my home state of Oklahoma. The practice has now been used on more than 1 million currently producing wells, 35,000 wells per year, without one confirmed case of groundwater contamination from these fracked formations. But don't take my word for it. Let's hear what the experts, the State regulators, have said:

Nick Tew, Alabama State Geologist & Oil and Gas Supervisor

"There have been no documented cases of drinking water contamination that have resulted from hydraulic fracturing operations to stimulate oil and gas wells in the State of Alabama."

Cathy Foerster, Commissioner of Alaska Oil and Gas Conservation Commission:

---

[*] This is an edited, reformatted and augmented version of Statement given by James M. Inhofe, before the Subcommittee on Water and Wildlife, Hearing on "Natural Gas Drilling: Public Health and Environmental Impacts" on Tuesday, April 12, 2011.

"There have been no verified cases of harm to ground water in the State of Alaska as a result of hydraulic fracturing."

Harold Fitch Director, Michigan Office of Geological Survey of the Department of Environmental Quality –

"There is no indication that hydraulic fracturing has ever caused damage to ground water or other resources in Michigan. In fact, the OGS has never received a complaint or allegation that hydraulic fracturing has impacted groundwater in any way."

Victor Carrillo, Chairman Railroad Commission of Texas –

"Though hydraulic fracturing has been used for over 60 years in Texas, our Railroad Commission records do not reflect a single documented surface or groundwater contamination case associated with hydraulic fracturing."

Fred Steece, Oil and Gas Supervisor of the South Dakota Department of Environment and Natural Resources –

"In the 41 years that I have supervised oil and gas exploration, production and development in South Dakota, no documented case of water well or aquifer damage by the fracking of oil or gas wells, has been brought to my attention. Nor am I aware of any such cases before my time."

And I have statements from 8 more oil and gas producing states that I would like to submit for the record. They all state that hydraulic fracturing does not contaminate ground water.

Now let me show you why this is the case. This chart illustrates a cross section of a typical well drilled in the Marcellus shale in southwest Pennsylvania. Do you see the small blue line at the top of the chart? That illustrates the ground water aquifer. In between that groundwater aquifer and the Marcellus shale are dozens of layers of solid rock – *more than a mile of it.* Let me say that again: there is more than a mile that separates the groundwater aquifer and the well.

See the small blue box at the top? That's a picture of the Empire State Building. For groundwater contamination to occur, frack fluids would have to migrate through 7,000 feet of that solid rock. Once again, that's about the same distance as from the West front of the Capitol all the way to the Washington Monument – of solid rock. That fluid migration can't happen and it doesn't happen.

Given these facts, what can possibly explain calls to regulate fracking from Washington, D.C.? It's simple: the Obama Administration wants to regulate fossil fuels out of existence. And they haven't been shy about it. Energy Secretary Steven Chu actually said, *"Somehow we have to figure out how to boost the price of gasoline to the levels in Europe."* For my colleagues who might not know, prices are $8 per gallon in Europe. Or consider Alan Krueger of the Treasury Department, who said, *"The Administration believes that it is no longer sufficient to address our nation's energy needs by finding more fossil fuels..."*

Mr. Krueger's belief is now reality. Gas at the pump is fast approaching $4.00 a gallon. Drilling in federal offshore waters is nearly non-existent. As for federal lands, the Western Energy Alliance recently reported that oil and gas leasing has dropped by 67 percent since 2005.

If you think these data points are bad, they will grow far worse under EPA's cap-and-trade agenda. As part of that agenda, the agency is manuevering to regulate hydraulic fracturing, a practice that has always been regulated by the states.

But testimony today will confirm that the states don't need EPA. The nation's immense shale deposits are predominantly located in states that effectively and efficiently regulate oil and gas. In states such as Pennsylvania, Arkansas, Oklahoma, Texas, Louisiana, West Virginia, Ohio, and North Dakota, a virtual boom in natural gas development is transforming America's energy security – due in no small measure to the *absence* of federal regulation.

For this reason, let's keep the states in charge of hydraulic fracturing, for the benefit of consumers, jobs, economic growth and expansion, and our nation's energy security.

In: Hydraulic Fracturing and Natural Gas Drilling ISBN: 978-1-61470-180-4
Editor: Aarik Schultz © 2012 Nova Science Publishers, Inc.

Chapter 6

TESTIMONY OF BOB PERCIASEPE, DEPUTY ADMINISTRATOR, U.S. ENVIRONMENTAL PROTECTION AGENCY, BEFORE THE SUBCOMMITTEE ON WATER AND WILDLIFE[*]

Good morning, Mr. Chairman and Members of the Subcommittee. I am pleased to be here today to discuss EPA's role in ensuring that public health and the environment are protected during natural gas extraction and production activities.

Natural gas can enhance our domestic energy options, reduce our dependence on foreign supplies, and serve as a bridge fuel to renewable energy sources. If produced responsibly, natural gas has the potential to improve air quality, stabilize energy prices, and provide greater certainty about future energy reserves.

While natural gas holds promise for an increased role in our energy future. EPA believes it is imperative that we access this resource in a way that protects human health and the environment.

Like you, we have heard the public's concerns about the safety of natural gas production and the potential impacts of shale gas development on American communities. As we listened to citizens at public meetings across the country last year, we heard the concerns many have for their families, their communities, the water, the air, and the land. We also heard from citizens who

---

[*] This is an edited, reformatted and augmented version of testimony given by Bob Perciasepe, Deputy Administrator, U.S. Environmental Protection Agency, before the Subcommittee on Water and Wildlife on April 12, 2011.

expressed how much their communities sorely need the income that could be gained from natural gas production.

We believe that this important resource can be – and must be – extracted responsibly, in a way that secures its promise for the benefit of all.

If improperly managed, natural gas extraction and production, including hydraulic fracturing, a technique for extracting natural gas, may potentially result in public health and environmental impacts at any time in the lifecycle of a well and its associated operations. Such impacts may include:

- stress on surface water and groundwater supplies from the withdrawal of large volumes of water used in drilling and hydraulic fracturing;
- potential contamination of drinking water aquifers resulting from faulty well construction and completion;
- compromised water quality due to challenges with managing and disposing of contaminated wastewaters, known as flowback and produced water, where contaminants could include organic chemicals, metals, salts and radionuclides; and
- impaired air quality from hazardous air pollutants such as benzene and the potent greenhouse gas methane.

Where we know problems exist, EPA will not hesitate to protect Americans whose health may be at risk, and we remain committed to working with state officials, who are on the front lines of permitting and regulating natural gas production activities.

EPA has an important role to play in ensuring environmental protection and in working with federal and state government partners to manage the benefits and risks of shale gas production. We must effectively address concerns about the consequences of shale gas development using the best science and technology. To this end, we are working in the following areas, among others, with all of our stakeholders, including other federal and state agencies, the oil and gas industry, and the public health community to evaluate and address the potential environmental issues related to shale gas.

RESEARCH

At the direction of Congress, EPA launched a study last year to understand the relationship between hydraulic fracturing and drinking water resources-- a study that has already engaged thousands of Americans across the country who are currently living in areas where hydraulic fracturing is taking place. When complete, this peer-reviewed research study will help us better understand potential impacts of gas extraction and production on drinking water resources and factors that may lead to human exposure and risks, while reducing scientific uncertainties about environmental impacts from those processes.

As part of this effort, EPA required information from nine oil and gas service companies conducting hydraulic fracturing regarding the chemical composition of the fracturing fluids they are injecting into the ground. EPA has conducted stakeholder outreach during development of the study plan, including opportunity for public comment. The draft study plan is being reviewed by EPA's Science Advisory Board and it included an opportunity for public comment. The initial study results are expected to be available in late 2012.

PROGRAMMATIC ACTIVITIES

While Congress specifically exempted selected oil and gas production activities from several environmental laws, a number of environmental protections continue to apply. For example, while the Energy Policy Act of 2005 excluded hydraulic fracturing for oil and gas production from permitting under the Safe Drinking Water Act's (SDWA) Underground Injection Control (UIC) Program, these activities are still regulated under the SDWA when diesel fuels are used in fracturing fluids. Also, flowback and produced water disposal through injection is still regulated under the SDWA.

EPA regulates waste waters from oil and gas wells under the Section 301(b) and 402 (a) of the Clean Water Act (CWA) when they are discharged into publicly owned treatment works (POTWs) and surface waters. To address public health concerns from air emissions, Clean Air Act (CAA) provisions for New Source Performance Standards and National Emission Standards for Hazardous Air Pollutants, and Mandatory Reporting of Green House Gases apply to energy extraction.

Under several of the principal environmental statutes we administer, EPA has a number of activities underway, which I would like to outline for you.

CWA AND SDWA

Under the CWA and SDWA, EPA works with states to ensure that gas extraction is carried out consistent with CWA and SDWA regulations to protect water and drinking water. For example, on March 16, 2011, we released shale gas extraction Frequently Asked Questions (FAQs) intended to serve as CWA guidance to state and federal permitting authorities within the Marcellus Shale region in addressing treatment and disposal of wastewater from shale gas extraction. The FAQs discuss the wastewater issues and pollutants associated with shale gas extraction and how they can be addressed under existing regulations. Relevant regulations that are discussed cover oil and gas extraction, centralized waste treatment, acceptance and notification requirements for publicly owned treatment works, pretreatment, and storm water. The FAQs should assist EPA and state personnel as they work with the regulated community to address shale gas extraction wastewater.

As part of its effluent guidelines planning process under CWA section 304(m), EPA is considering whether to initiate a rulemaking to revise these regulations to address coal bed methane extraction flowback waters. Also, in response to public comment, EPA is considering how best to address shale gas extraction wastewater discharges to POTWs under the CWA. Similarly, under SDWA's UIC program, EPA is working expeditiously to ensure the SDWA programmatic requirements related to hydraulic fracturing when using diesel fuels are implemented appropriately. We are coordinating with our state and Tribal co-regulators to ensure proper management of flowback and produced water disposed of via underground injection.

CLEAN AIR ACT (CAA)

A range of Clean Air Act (CAA) provisions apply to the oil and gas sector. Under the CAA New Source Performance Standards and National Emission Standards for Hazardous Air Pollutants programs, EPA must review, and propose amendments, where warranted based on the reviews, to existing CAA regulations for oil and natural gas production, natural gas processing plants, and natural gas transmission and storage.

Subpart W of EPA's Greenhouse Gas Reporting Program requires companies to begin collecting methane emissions data as of January 1, 2011, and report the emissions for each year starting in 2011 by March 31 of the following year. This rule will increase the understanding of the location and magnitude of significant methane and other GHG emissions sources in the petroleum and natural gas industry. Increased information will help companies improve the efficiency of their operations and will result in cross-cutting benefits for public health, domestic energy supply, industrial efficiency and safety, and revenue generation through methane recovery and other emissions reduction efforts.

EPA's regulatory efforts will complement our Natural Gas STAR Program, a flexible, voluntary partnership between EPA and oil and natural gas operating companies which encourages companies both in the United States and internationally to adopt proven, cost-effective technologies and practices that improve operational efficiency and reduce methane emissions. Beginning in 1993, this successful voluntary program now has over 130 partner companies. Together we have identified over 80 technologies and practices that can cost-effectively reduce methane emissions from the oil and natural gas sector. (A list of these technologies is posted on the EPA Natural Gas STAR website: (http://www.epa.gov/gasstar/tools/ recommended.html). These include methane emission reduction activities applicable to the largest natural gas industry methane sources: gas well liquid unloading, unconventional gas well completions and workovers, pneumatic devices, reciprocating and centrifugal compressors, crude oil and condensate tank vents and general fugitive emissions. Natural Gas STAR partners reported domestic emissions reductions of 86 Bcf, worth over $344 million, in 2009 and over 900 Bcf, worth over $4.4 billion, over the life of the program.

ENFORCEMENT

Under each of these statutes I just described, EPA has tools for enforcement and investigation to assure compliance with applicable requirements, and the Agency provides support and technical assistance to state programs, which have front-line responsibility for oversight of oil and gas drilling operations.

In April 2010 EPA announced an enforcement initiative for energy extraction sector for fiscal years 2011 through 2013. The initiative has twin goals: to take appropriate enforcement action where adverse impacts on air and water from energy extraction activities threaten human health, and to incorporate broad, company-wide injunctive relief in gas-related enforcement actions as a means of reducing human health and environmental impacts of the industry.

EPA retains authority to respond to imminent and substantial endangerments to public health or the environment under several statutes, including the Safe Drinking Water Act, the Clean Air Act, the Clean Water Act, the Resource Conservation and Recovery Act, the Toxic Substances Control Act, and the Comprehensive Environmental Response, Compensation, and Liability Act.

CONCLUSION

In conclusion, EPA is committed to using its authorities, consistent with the law and best available science, to protect communities across the nation from impacts to water quality, human health, and environment associated with natural gas production activities. We also commit to coordinating with our federal, state, and local partners as we move forward. By helping manage environmental impacts and address public concerns, natural gas production can proceed in a responsible manner, which protects public health and enhances our domestic energy options. We believe that by doing so, as a nation, together we can establish a sound framework that allows for the safe and responsible development of a significant domestic energy resource whose use brings a range of other important national security, environmental and climate benefits.

Chapter 7

WRITTEN TESTIMONY OF CONRAD DANIEL VOLZ, GRADUATE SCHOOL OF PUBLIC HEALTH, UNIVERSITY OF PITTSBURGH, BEFORE THE SUBCOMMITTEE ON WATER AND WILDLIFE, HEARING ON "NATURAL GAS DRILLING: PUBLIC HEALTH AND ENVIRONMENTAL IMPACTS"[*]

Thank you for the opportunity to testify this morning at the Joint Hearing on Natural Gas Drilling, Public Health and Environmental Impacts. Unconventional gas extraction in deep shale deposits presents considerable risks to public health and safety as well as to environmental resources, particularly water quality and aquatic organisms. My testimony today will cover three critical public health and environmental policy areas related to unconventional natural gas production.

First is the unregulated siting of natural gas wells in areas of high population density, and near schools and critical infrastructure. Unconventional gas extraction wells are highly industrialized operations that have public health preparedness risks of catastrophic blowout, explosion and fire. Any of these incidents can create an Immediately Dangerous to Life and

---

[*] This is an edited, reformatted and augmented version of testimony given by Conrad Daniel Volz, Graduate School of Public Health, University of Pittsburgh, before the Subcommittee on Water and Wildlife, Hearing on "Natural Gas Drilling: Public Health and Environmental Impacts" on April 12, 2011.

Health (IDLH) condition for adults or children in close physical proximity. The unregulated siting of unconventional natural gas extraction wells and production facilities in residential neighborhoods and near critical infrastructure is unwise preparedness policy, especially in light of federal and state efforts to reduce risk from terror attacks on USA citizens and critical infrastructure.

Secondly, the higher rates and differential patterns of oil and gas act violations from Marcellus Shale gas extraction operations, as compared to conventional oil and gas wells, suggests a much greater impact to drinking water and aquatic resources. Marcellus Shale gas extraction wells have between 1.5 to 4 times more violations than their conventional well counterparts per offending well, including more serious violations and violations that have a direct impact on water quality and aquatic resources. Marcellus Shale gas extraction wells are more likely to have violations for:

- Failures to minimize accelerated erosion, implement erosion and sedimentation plans, and/or maintain erosion and sedimentation controls.
- Discharge of pollution to waters of the Commonwealth of Pennsylvania.
- General violations of the Clean Streams Law.
- Failure to properly store, transport, process or dispose of a residual waste and -
- Failures to adequately construct or maintain impoundments holding gas extraction flowback fluids containing toxic contaminants.

The third problem public health and environmental policy area to be addressed is the disposal of gas extraction flowback fluids, carrying a plethora of toxic elements and chemicals, in inefficient "brine" treatment facilities and Publicly Owned Treatment Works (POTW's) [commonly called sewage treatment plants], which discharge effluent into surface water sources. Studies of the effluent from a commercial facility in Pennsylvania that treats fluids only from gas and oil operations shows discharge of 9 pollutants in excess of nationally recognized human and/or aquatic health standards into a nearby stream. The contaminants include:

- Barium, found in effluent over 8 times its minimum risk level (MRL) in drinking water to children and 27 times its EPA consumption concentrations for fish and "fish plus water".

- Stable Strontium, found in effluent 43.29, 51.68 and 97.90 times the drinking water MRL's for intermediate exposures for adult men, adult women, and children, respectively. Strontium levels found in effluent were 29,811 times the reporting limit in the plants NPDES permit.
- Bromide, which forms mixed chloro-bromo byproducts in water treatment facilities that have been linked to cancer and other health problems were found in effluent at 10,688 times the levels generally found acceptable as a background in surface water.
- Benzene, a known carcinogen, is present in effluent water at over 2 times its drinking water standard, over 6 times its EPA consumption criteria, and 1.5 times the drinking water MRL for chronic exposure for children.
- 2-butoxyethanol (2-BE), a glycol ether and used as an antifoaming and anticorrosion agent in slick-water formulations for Marcellus Shale gas extraction was found in effluent water at 24.48, 29.21, and 55.14 times the drinking water MRL's for intermediate exposure to adult males, adult females, and children, respectively –based on hepatic health effects.
- Chlorides, the concentration of chlorides in the effluent was 138 and 511 times the EPA maximum and continuous concentration criteria set for the health of aquatic organisms, respectively.

Due to time constraints I will not cover impacts to air quality, although I wish to go on record that these impacts could be significant, due to release of hazardous air pollutants from 10's of thousands of projected natural gas wells, with the subsequent formation of ozone; areas of Maryland, Pennsylvania, Ohio, New York, and New Jersey are already in EPA nonattainment status for ozone exposure.

Potential "Immediately Dangerous to Life and Health" (IDLH) Conditions from Unregulated Siting of Unconventional Gas Extraction Wells

Unconventional gas extraction wells are highly industrialized operations that have attendant risks of catastrophic blowout, explosion and fire. The actualization of any of these incidents creates an IDLH condition for adults and children in close proximity to these wells from any blast or fires, the displacement of oxygen by methane, exposure to waterborne contaminants, and from inhalation of pyrolysis products of burning condensate, liners and/or production equipment. Over the past 2 years, within a 3 hour drive of

Pittsburgh PA, there has been one catastrophic blowout, one explosion and fire due to ignition of methane from an underground coal mine, and two fires (one at a multi-well site in production near Avella PA and one at a site being brought into production in Hopewell Township, PA).

If we use the figure of 1831 drilled wells in the State of Pennsylvania from 2007 to September of 2010, which is an overestimate of the wells drilled in a three hour drive of Pittsburgh PA and use this as the denominator, and use 4 incidents as the numerator we obtain an order of magnitude estimate of the probability for IDLH conditions at these wells of 0.002. Using this figure and based on estimates of the predicted number of wells to be drilled over the next 10 years of 25,000 wells- there could be as many as 50 wells that create IDLH conditions due to blowout, blast and/or fire. What is disturbing, in this era of spending billions of dollars to reduce risk from terror attacks on USA citizens and critical infrastructure, is that we are allowing these gas extraction wells to be sited in a largely unregulated fashion in close proximity to homes and critical infrastructure including schools, and in densely populated regions of Pennsylvania, Ohio, and West Virginia.

The well publicized and documented Marcellus Shale blowout in Clearfield County PA, due in part to failure of the operator to properly test the Blow-Out Preventers (BOPs) prior to use and to conduct the BOP test in a proper manner, resulted in the immediate evacuation of all residents within one mile of the drill site. Luckily the impacted area was largely state forest land with no population proximal to the drill site and very diffuse population density. CHEC has done projections to show impacts of such a blowout in a more densely populated area south of Pittsburgh PA- Peters Township, Washington County PA, where gas leases are currently being signed. If the blowout had occurred in the centroid of this township approximately 1,928 adults and children would need to be evacuated as well as up to 5 school complexes. Local emergency response personnel are not properly trained or adequately equipped to handle these type incidents nor is there a gas extraction specific planning mechanism for such large population displacements.

Patterns of Oil and Gas Violations from Marcellus Shale Gas Extraction Operations in Pennsylvania and General Threats to Water Resources

CHEC analyzed the number of Oil and Gas Act violations by well type in Pennsylvania over the period from January 1, 2007 to September 30, 2010 and found that Marcellus Shale gas extraction wells have between 1.5 to 4 times more violations than their conventional well counterparts per offending well (this is dependant on the denominator of total wells drilled which is difficult to ascertain for conventional oil and gas wells due to drilling for over 100 years). These include more serious violations and violations that potentially have a more direct impact on water quality and aquatic resources. Between January 1, 2007 and September 30, 2010, horizontal Marcellus wells had 3.75 violations per offending well, while vertical Marcellus wells had 2.99 violations per offending well, resulting in a rate of 3.51 violations per offending well for all Marcellus wells. Conventional non-Marcellus oil and gas wells had violations per offending well of 2.38.

In 2010, 451 distinct Marcellus Shale gas extraction wells in Pennsylvania were cited for violations of the Oil and Gas Act by the Pennsylvania Department of Environmental

Protection (DEP). There were 1544 total violations resulting in a mean violations rate per offending well of 3.42. Of these 1544 total Marcellus violations; 111 violations were for failure to minimize accelerated erosion, implement erosion and sedimentation plans, and/or maintain erosion and sedimentation controls and/or failure to stabilize the site until total site restoration under OGA Section 206(c)(d); 105 violations were for discharge of pollution to waters of the Commonwealth; 106 violations were general violations of the Clean Streams Law; 68 violations were for failure to properly store, transport, process or dispose of a residual waste; and 116 violations were issued for impoundment problems including failure to maintain a 2 foot freeboard, and impoundment not structurally sound or impermeable.

These patterns of violations of the Pennsylvania Oil and Gas Act by Marcellus Shale gas operators support my contention that development of natural gas from the Marcellus Shale has the potential to result in substantial adverse effects on water quality, the environment and public health. Ground-surface disturbances associated with well drilling, including site clearing, and the construction of access roads, drill pads and impoundments, can produce impacts associated with stormwater, erosion and sedimentation of surface waterways, which in turn may lead to higher levels of water turbidity, total dissolved solids, conductivity and salinity. In addition to the impacts associated with surface activities are those associated with deep well drilling. Wells drilled to depths of 5,000 to 8,000 feet to reach the Marcellus formation (and also the Utica Shale formation) create pathways for the migration of naturally-occurring contaminants into usable quality aquifers, and involve the disposition on the surface of drill cuttings and formation waters that also may contaminate ground and surface water. Contaminants associated with natural gas drilling in the Marcellus include toxic heavy metals and elements, organic compounds, radionuclides and acid producing sulfide minerals, and natural gases and sulfide producing gases, which can threaten surface and groundwater sources.

Disposal of Oil and Gas Flowback Fluids in Inefficient "Brine" Treatment Facilities and Publicly Owned Treatment Works (POTW's) that Discharge into Surface Water; Potentially Exposed Populations and Regional Significance

Hydraulic fracturing (HF) of shale gas deposits uses considerable masses of chemicals, for a variety of purposes to open and keep open pathways through which natural gas, oil and other production gases and liquids can flow to the wellhead. HF, also known as slick-water fracturing, introduces large volumes of amended water at high pressure into the gas bearing shale where it is in close contact with formation materials that are enriched in organic compounds, heavy metals and other elements, salts and radionuclides. Typically, about 1 million gallons and from 3 - 5 million gallons of amended water are needed to fracture a vertical well and horizontal well, respectively (Hayes, 2009). Fluids recovered from these wells can represent from 25% to 100% of the injected solution and are called "flowback" or "produced" water depending on the time period of their return. Flowback and produced water contain high levels of total dissolved solids, chloride, heavy metals and elements as well as enriched levels of organic chemicals, bromide and radionuclides – in addition to the frac chemicals used to make the water slick-water. Levels of shale origin contaminants in flowback water generally increase with increasing time in contact with formation materials.

This oil and gas fluid waste is generally held in temporary open-air impoundment(s) near the well site or occasionally in large sealed containers. Additionally, oil and gas waste fluids accumulate in condenser tanks located on producing well pads, which must be drained regularly. Currently, flowback water is either taken for disposal to a POTW (sewage treatment plant), or a Brine Treatment Facility, both of which discharge effluent directly to surface water sources. The waste fluids may also be recycled for reuse (on-site or off-site at treatment facilities), or injected into Class II underground wells.

The relative volumes of flowback and condensate entering each end-point alternative described above are currently the subject of much heated debate, the unraveling of which is well beyond the scope of my testimony. It is sufficient to note that large volumes of oil and gas wastewater are disposed of in POTW's and brine treatment facilities that discharge effluent directly into surface water. The PA Brine Treatment, Josephine Facility received 15,728,242 gallons of Marcellus Shale gas extraction wastewater for treatment and effluent discharge into Blacklick Creek, Indiana County in the last half of 2010. The Clairton POTW received and disposed of 53,473 gallons of Marcellus Shale wastewater in the last half of 2010, which is ultimately discharged into the Monongahela River. CHEC has identified at least 10 facilities that discharged effluent into the Monongahela River drainage in 2010-2011 in Pennsylvania alone; if all these facilities are accepting flowback fluids at their permitted rate then 824,000 pounds of total dissolved solids and 15,000 pounds of barium could enter the watershed from these operations daily.

There is considerable scientific inquiry and even controversy regarding the potential of vertical or horizontal fracturing of shale gas reservoirs to contaminate shallow or confined groundwater aquifers, and thus expose municipal or private well water users to chemicals used in the hydrofracturing process and/or contaminants in the formation materials. However, when Marcellus Shale flowback and produced fluids are disposed of in POTW's or inefficient brine treatment facilities discharging into surface water, the fate and transport pathways to expose human and aquatic receptors are well described for most of the contaminants potentially in effluent discharge water and known to be in flowback and other oil and gas wastewater. Contaminants untreated by the facility and discharged into surface water will move in the water through advective and fickian processes downstream, be deposited and transferred into sediments and pore water, bioaccumulate in aquatic receptors and terrestrial animals that feed on them according to their species specific bioaccumulation factors, be transported to groundwater, and/or be volatilized to air dependent on their Henry's Law constants. Direct and complete human and ecological exposure pathways via ingestion, dermal absorption and inhalation (gill transfer in fish) can be demonstrated for different classes of elements, and compounds in the wastewater, constituting a potential exposure threat to recreationalists, private well water users and municipal drinking water users.

Case Example; Concentrations of Contaminants in Effluent Water from Pennsylvania Brine Treatment Facility, Josephine Facility (PBT-JF)

The Center for Healthy Environments and Communities (CHEC) of the Graduate School of Public Health, University of Pittsburgh, conducted sampling of wastewater as it was discharged into Blacklick Creek, Indiana County, Pennsylvania from the PBT-JF on December 10, 2010. Samples were taken at 3-hour intervals over the course of one 24- hour period. The concentrations of analyzed contaminants in this effluent of primary environmental public health importance, which may also stress aquatic life, include: barium (Ba) [mean, 27.3 ppm; maximum, 37.0 ppm]; bromides (Br) [mean, 1068.8 ppm; maximum, 1100.0 ppm; strontium (Sr) [mean, 2983.1 ppm, maximum 3120.0 ppm]; benzene [mean 0.012 ppm; maximum 0.013 ppm] and 2-butoxyethanol (2-BE) [mean 59ppm; maximum 66 ppm]. Contaminant concentrations of ecological and secondary drinking water importance include: chlorides (Cl) [mean 117,625 ppm, maximum 125,000 ppm]; magnesium (Mg) [mean 1247.5 ppm; maximum 1300.0 ppm]; total dissolved solids (TDS) [mean 186,625 ppm; maximum 190,000 ppm]; sulfate (SO4) [mean 560 ppm; maximum 585 ppm], and pH [mean 9.58 units; maximum 10 units].

Levels of contaminants in effluent from the PBT-JF were interpreted according to comparisons with applicable federal and state standards and recommended guidelines for both human and aquatic health. Barium had a mean concentration in effluent of 27.3 ppm (maximum of 37 ppm); this is approximately 14 times the United States Environmental Protection Agency (EPA) maximum concentration limit (MCL) of Ba in drinking water of 2 ppm. The EPA consumption concentrations 'water and organism' and 'organism alone' for barium are both 1 ppm. The levels of barium in the effluent are over 27 times these consumption concentrations. The U.S. EPA criteria maximum concentration (CMC) and the EPA criteria continuous concentration (CCC), both for protection of aquatic health, are 21 ppm and 4.1 ppm, respectively; the mean level of barium in effluent exceeds these criteria by 1.3 and 6.7 times, respectively. The mean concentration of barium in PBT-JF effluent water (27.3 ppm) is 3.96, 4.73, and 8.98 times the ATSDR derived drinking water minimum risk level (MRL) for intermediate and chronic exposures for adult men, and women, and children, respectively.

The EPA (ATSDR ascribed) recommends that drinking water levels of stable strontium should not be more than 4 milligrams per liter of water (4 mg/L), Sr levels in PBT-JF effluent are 746 times this recommended level. The strontium ATSDR MRL for oral route, intermediate exposure is 2 mg/kg of body mass/day, for musculoskeletal endpoints. The derived minimum risk levels for strontium in drinking water for intermediate exposure for adult men, adult women, and children are 68.87 mg/L/day, 57.67 mg/L/day, and 30.45 mg/L/day, respectively. The mean concentration of strontium in PBTJosephine effluent water (2,981.1 ppm) is 43.29, 51.68 and 97.90 times the derived strontium drinking water MRL's for intermediate exposures for adult men, adult women, and children, respectively. Strontium is not listed on the PBT-JF, NPDES permit but the facility is required to notify the PA DEP if they routinely discharge 100 ppb of a toxic pollutant or nonroutinely discharge 500 ppb of a toxic pollutant. The mean concentration of Sr in effluent water of 2,981.1 ppm is 29,811 and 5,962 times the lower and upper notification levels required by the PA DEP NPDES permit, respectively. . Searches of the PA DEP file for December, 2010, shows no such notification to the DEP.

Bromide in water is of concern because of its ability to form brominated analogs of drinking water disinfection by-products (DBP). Specifically, bromide can be involved in reactions between chlorine and naturally occurring organic matter in drinking-water, forming brominated and mixed chloro-bromo byproducts, such as trihalomethanes or halogenated acetic acids. Several DBPs have been linked to cancer in laboratory animals, and as a result the U.S. EPA has regulated some DBP's. There is general agreement that bromide levels in freshwater sources be kept below about 100 ppb (.1 ppm) so that formation of brominated DBP's are minimized, therefore regulatory authorities and water treatment plant operators become concerned when there are sources of bromides in a surface system adding to this level. The PBT-JF discharged effluent into Blacklick Creek with a measured mean concentration of bromide of 1,068.8 ppm, which is 1,068,800 ppb. This is 10,688 times the 100 ppb level at which authorities become concerned. Bromide is not listed on the PBT-JF NPDES permit, but the facility is required to notify the PA DEP if they routinely discharge 100 ppb of a toxic pollutant or nonroutinely discharge 500 ppb of a toxic pollutant. The mean concentration of Br in effluent water 1,068.8 ppm is 10,688 and 2,138 times the lower and upper notification levels required by the PA DEP NPDES permit, respectively. Searches of the PA DEP file for December, 2010, shows no such notification to the DEP.

The mean level of benzene, a known carcinogen, in outfall effluent from PBT-JF was 0.012 ppm or 12 ppb. The drinking water MCL for benzene is 5 ppb, thus effluent levels were above twice the drinking water MCL. The EPA consumption, water and organism risk level for benzene is 2.2 ppb in water, the mean level of benzene in PBT-Josephine effluent water is almost 6X this criteria; the organism only risk level for benzene is 50 ppb in water, the mean level of benzene in effluent water is 24% of this guideline. The benzene ATSDR MRL for oral route, chronic exposure is 0.0005 mg/kg of body mass/day, for immunological endpoints. The derived minimum risk levels for benzene in drinking water for chronic exposure for adult men, adult women, and children are 0.017 mg/L/day, 0.014 mg/L/day, and 0.008 mg/L/day, respectively. The mean concentration of benzene in PBT-Josephine effluent water (0.012 ppm) is 70% of, 86% of, and 1.5 times the derived chronic drinking water MRL for benzene for adult men, adult women, and children, respectively.

2-butoxyethanol (2-BE) is a glycol ether and is used as an antifoaming and anti-corrosion agent, as well as an emulsifier in slick-water formulations for Marcellus Shale gas extraction. The mean and maximum levels of 2-BE found in the PBT – JF effluent were 59 ppm and 66 ppm, respectively. The 2-BE ATSDR MRL for oral route, acute exposures is 0.4mg/kg/day based on hematological effects, with an uncertainty factor of 90; the 2B-E MRL for oral route, intermediate exposure is 0.07 mg/kg/day and it is based on hepatic health endpoints with an uncertainty factor of 1000. The derived minimum risk levels for 2-BE in drinking water for acute exposure for adult men, adult women, and children are 13.77 mg/L/day, 11.53 mg/L/day, and 6.09 mg/L/day, respectively; the derived MRL's for 2-BE in drinking water for intermediate exposure for adult men, adult women, and children are 2.41 mg/L/day, 2.02 mg/L/day, and 1.07 mg/L/day, respectively. The mean concentration of 2-BE in PBT-JF effluent water (59 ppm) is; 4.28, 5.12, and 9.69 times the derived 2-BE drinking water MRL's for acute exposure to adult males, adult females, and children, respectively; and 24.48, 29.21, and 55.14 times the derived 2-BE drinking water MRL's for intermediate exposure to adult males, adult females, and children, respectively. 2-BE is not listed on the PBT-JF NPDES permit, but the facility is required to notify the PA DEP if they routinely discharge 100 ppb of a toxic pollutant or nonroutinely discharge 500 ppb of a toxic pollutant. The mean concentration of 2-BE in effluent water is 590 and 118 times the lower and upper notification levels, required by the PA DEP NPDES permit, respectively. Searches of the PA DEP file for December, 2010 show no such notification to the DEP.

Contaminants with secondary MCL's (SMCL) and aquatic receptor effects that were measured in the PBT-JF effluent include magnesium, manganese, chlorides, sulfates, and total dissolved solids (TDS). Magnesium was found in the effluent with a mean concentration of 1,247.5 mg/L, which is 24,950 times the EPA Mg SMCL of .05 mg/L. The mean concentration of Manganese in the effluent was .08 mg/L, and the SMCL for Manganese concentration in drinking water is .05 mg/L, which is 62.5% lower than the concentration in the effluent. The mean concentration of chlorides in the sample analysis was 117,625 mg/L, which is 470.5 times the SMCL for chlorides in drinking water of 250 mg/L. To protect aquatic communities, the criteria maximum concentration (CMC) for chlorides in surface water is 860 mg/L, and the criteria continuous concentration (CCC) for chlorides in surface water is 230 mg/L. The mean concentration of chlorides measured in samples was 138 times the CMC and 511 times the CCC. The mean concentration of sulfates in the sample analysis was 560 mg/L, 2.2 times the SMCL for sulfates in drinking water (250 mg/L). The SMCL for total dissolved solids (TDS) in drinking water is 500 mg/L, and the mean concentration of TDS measured in samples was 186,625 mg/L, 373 times the SMCL.

MASSES OF CONTAMINANTS ENTERING BLACKLICK CREEK

CHEC has information from the Pennsylvania, Department of Environmental Protection (DEP) that the PBT – JF treated 15,728,241 gallons of oil and gas wastewater in the 6 month period from July 1, 2010 to December 31, 2010. Using this figure as the amount of effluent wastewater exiting the Josephine outfall and using the mean level of each contaminant found in the effluent over the sampling period of the study, the masses of contaminants with important human and ecological consequences discharged from the PBT-JF into Blacklick Creek in the last 6 months of 2010 are projected to be: barium-1627 kg (3588 pounds); strontium -177,712 kg (391,856 pounds; 196 tons); bromides-63,708 kg (140,476 pounds; 70.2 tons); chloride – 7,011,631 kg (15,460,646 pounds; 7,730 tons); sulfate – 33,382 kg (73,607 pounds; 36.8 tons); 2 BE– 3517 kg (7,755 pounds; 3.88 tons); and total dissolved solids – 11,124,733 kg (24,530,036 pounds; 12,265 tons).

POTENTIALLY EXPOSED POPULATIONS

Recreationalists are at risk of being exposed to outfall contaminants through ingestion, inhalation and through dermal exposure. The outfall of the BBT-JF is easily accessible to users of nearby rails-to-trails pathways, and there are indications that anglers frequent the area. Additionally, children wade and swim in the creek during warmer weather, and regional watershed websites indicate that paddlers use the creek for canoeing and kayaking. 2 BE released into Blacklick Creek may be ingested by swimmers in the creek. This pollutant can become airborne and present an inhalation hazard to anglers, swimmers and boaters. It is also taken in to the body via dermal absorption. Anglers catching and eating fish from upstream or downstream of the effluent outfall are at risk for exposure to multiple contaminants that were sampled in this study. CHEC has developed maps showing numerous private water wells in the immediate vicinity of Blacklick Creek downstream from the effluent discharge. Private well water users are at risk of exposure to contaminants in effluent being released into Blacklick Creek because these private wells may capture water from the creek when the well pump rate is sufficiently high. High pump rates can occur especially during peak usage by residents. The first identified municipal drinking water intake downstream of this discharge is at Freeport, Pennsylvania on the Allegheny River. Populations served by the Freeport authority and water authorities downstream of Freeport are at potential risk for exposure to contaminants identified in effluent, as well as other contaminants in Marcellus Shale flowback water that were not sampled for in this study.

Implications of Effluent Discharge from the PBT – Josephine Facility Discharge For Exposures to Other Contaminants Known to be Present in Marcellus Shale Flowback Fluids and a Regional Appreciation of These Results

Of particular environmental public health significance is that Marcellus Shale flowback water contains other contaminants, in addition to those analyzed for in this study, which have health consequences if ingested, inhaled, and/or absorbed through the skin. While we make no statements regarding the presence of other contaminants in this effluent water being discharged into Blacklick Creek, it is imperative that additional testing be conducted immediately by federal and state health and enforcement agencies to determine if other contaminants of public health significance are entering this watershed. Additionally, oil and gas wastewater and Marcellus shale flowback fluids are being disposed of in "brine treatment" facilities and at POTW's throughout the Commonwealth of Pennsylvania and in Ohio, Maryland, West Virginia, and New York. The ramifications of disposal of large quantities of oil and gas wastewater through ineffectual brine treatment plants and POTW's needs further evaluation throughout the region to determine its impact on stream and river systems and public drinking water supplies, as well as to recreationalists and private well water users.

Local and Regional Public Health and Environmental Recommendations Based on PBTJF Results

- The Pennsylvania Brine Treatment – Josephine Facility is discharging up to 60 ppm of 2-BE into Blacklick Creek, which is not listed in its discharge permit. Operations at this plant should be halted until all contaminants in accepted oil and gas fluids are known and it can be determined if the treatment processes used at the plant effectively remove these contaminants from the fluids being treated, so that effluent discharge concentrations of contaminants are consistent with human and aquatic health standards, guidelines and criteria. This recommendation should extended to other treatment plants and POTW's accepting Marcellus Shale flowback fluids in this drainage.
- All approaches to the effluent discharge area and a reasonable distance downstream (at least 100 meters) from stream-side and land-side should be posted with warning signs. These signs should discourage any use of and/or contact with stream water.
- An advisory should be issued to all anglers that fish taken from this stream, both up and down stream, may be contaminated and discouraging fish take and of course consumption.
- Studies to determine the levels of all potential Marcellus Shale flowback fluid contaminants in downstream water, sediments and pore water should be undertaken immediately. These should include sampling upstream of the effluent discharge point and at short, intermediate and longer distances downstream from the effluent discharge point. The number of samples taken (n) of surface water, sediments and pore water upstream and at the various distances downstream should be sufficient so that statistically significant differences of contaminant concentrations can be inferred. CHEC took additional samples of effluent and performed both up and downstream transect sampling on April 1st and 2nd, 2011- these samples are now being analyzed for an expanded list of chemicals including antimony, radium radionuclides, phenols and derivatives, polynuclear aromatic hydrocarbons (PAH's), phthalates, and total petroleum hydrocarbons
- Residential and other private well water users downstream of the effluent outfall of the PBT-Josephine Facility should be advised that there may be contaminants in their well water and discouraged from using it for drinking, cooking or bathing. Well water from wells in close proximity to Blacklick Creek should be tested to assure that contaminants in Marcellus Shale flowback fluids and other oil and gas waste fluids are not present in concentrations that may affect human health.

- Municipal water authorities downstream of this outfall should be notified of the contaminants found in effluent from the PBT-Josephine Facility, of other possible contaminants in Marcellus Shale flowback fluids and oil and gas wastewater, and that there are other treatment facilities and POTW's in the Blacklick, Conemaugh, and Kiskikiminetas drainages that accept and discharge oil and gas waste fluids into surface water. They should also be notified that landfill facilities in the drainage accept solid wastes produced from these treatment facilities. Downstream municipal water authorities should test raw unfinished intake water and finished drinking water for identified contaminants in effluent from the PBT- Josephine Facility, and other contaminants known to be present in Marcellus Shale flowback fluids and oil and gas wastewater.
- All municipal water authorities at reasonable distances downstream of "brine treatment" and POTW's accepting Marcellus Shale flowback fluids and other oil and gas wastewater in the region extending eastward across Ohio, Pennsylvania and West Virginia and New York should be notified of these results. It is important that they initiate sampling of raw, unfinished inflow water and finished drinking water immediately to insure that their systems are capable of handling all potential contaminants, without breakthrough above specific drinking water MCL's.
- The PA DEP and other states and federal regulatory authorities, including the Susquehanna River Basin Commission (SRBC) and the Delaware River Basin Commission (DRBC) should immediately review all surface water discharge permits granted to brine treatment facilities and POTW's that accept Marcellus Shale flowback fluids and oil and gas wastewater, to insure that 2-BE concentrations being discharged are below all applicable standard, guidelines and criteria. This review should be informed by results of this report but should be extended to all known contaminants in flowback and other oil and gas wastewater.

Chapter 8

TESTIMONY OF JOHN W. UBINGER, JR., SENIOR VICE PRESIDENT, PENNSYLVANIA ENVIRONMENTAL COUNCIL, BEFORE THE COMMITTEE ON ENVIRONMENT AND PUBLIC WORKS, HEARING ON "MARCELLUS SHALE DEVELOPMENT IN PENNSYLVANIA"[*]

INTRODUCTION

The Marcellus Shale is one of the largest unconventional on-shore gas deposits in the world. Estimated at between 250-500 trillion cubic feet of gas deep underground, the Marcellus Shale represents a natural gas supply that could meet America's energy needs for the next 50-80 years or more.

It is widely considered that the Marcellus Shale play offers an abundant fuel to help bridge the gap between today's energy portfolio and a future supply that reflects both a reduced carbon footprint and reduced dependence on foreign sources of energy.

There is both a national security interest as well as a private sector interest in this extraordinary resource, setting the stage for a truly unique opportunity

---

[*] This is an edited, reformatted and augmented version of testimony given by John W. Ubinger, Jr., Senior Vice President, Pennsylvania Environmental Council, before the Committee on Environment and Public Works, Hearing on "Marcellus Shale Development in Pennsylvania" on April 12, 2011.

for economic development, energy security, private sector profitability and public revenue generation.

The promise of this new industry comes at a critical time in our history, when bridge fuels to the future are desperately needed to help reduce our dependence on foreign sources of oil. At the same time, the recession has created a state budget in need of new sources of revenue. Additionally the prospect of new job creation from natural gas development and new industries attracted to Pennsylvania by a reliable natural gas supply comes at a critical time in our Commonwealth, as the unemployment rate in one-quarter of Pennsylvania remains at over ten percent.

Without question, the Marcellus Shale is a once-in-a-lifetime situation, and one that is already underway. The size and potential of the Marcellus has set off a "Pennsylvania Gas Rush," analogous to the California Gold Rush, the Texas oil boom and the discovery of oil on Alaska's North Slope.

But for all the excitement and promise of new economic opportunity, there are striking similarities to other energy resource development booms in Pennsylvania's history. Indeed, Pennsylvania has paid a very heavy price for the development of timber, coal and other extracted resources. That price has even yet to be fully paid and is evidenced by over 5,000 miles of polluted waterways, thousands of abandoned mines and oil and gas wells, decaying infrastructure, and economic devastation caused by poor planning and a short-sighted thirst for growth decades ago.

We need to learn from the mistakes of the past, apply them to the development of the Marcellus Shale, and make sure that we do everything possible to create a sustainable, thriving, and successful Pennsylvania Marcellus Shale economy that does not leave an environmental burden to future generations.

There is widespread agreement that "business as usual" in Marcellus Shale natural gas operations, as well as its current regulatory oversight, is not equal to the scale and scope of this development, and that simply applying conventional solutions to these significant challenges will result in adverse consequences to all stakeholders in the process.

The Pennsylvania Environmental Council (PEC) has taken the lead in bringing together representatives of communities, the natural gas development industry, government and environmental interest organizations in the spirit of finding a strategy that all sides can agree will accomplish three fundamental goals:

1) Enable the gas industry to prosper in the successful development of the Marcellus Shale and other deep shale gas plays in Pennsylvania;
2) Ensure that Pennsylvania benefits from the success of this industry, while preventing long-term costs; and
3) Protect people and the environment from adverse effects that result from the expansion of drilling operations.

To begin this dialogue, PEC held the Pennsylvania Marcellus Shale Policy Conference in Pittsburgh on May 3-4, 2010. The goal of this forum was to identify the key issues, challenges and opportunities in the effective and sustainable development of a Marcellus Shale gas industry in Pennsylvania.

From this conference, PEC produced a detailed report, "Developing the Marcellus Shale: Environmental Policy and Planning Recommendations for the Development of the Marcellus Shale Play in Pennsylvania" in July 2010.[1] This report represents PEC's findings and conclusions from that public dialogue, allowing for further research and analysis.

The policy recommendations in the report were offered to serve as the basis for new legislation and regulation designed to identify a framework whereby this vast natural resource can be developed for the benefit of America's energy portfolio, the private sector, and key stakeholders, while at the same time safeguarding the future prosperity of communities and the natural environment in Pennsylvania for current and future generations to come.

Since that time, PEC has been actively engaged in meetings with the members of the gas industry, state regulators, local governments and other environmental organizations aimed at bringing consensus to this debate around PEC's fundamental objectives. We were selected to serve as one of four environmental interest organizations on Pennsylvania Gov. Corbett's Marcellus Shale Commission in recognition of the leadership role PEC has played in seeking to bring about effective legislation and regulatory changes that give state government the resources and authority to effectively safeguard the people and environment of Pennsylvania with laws that are relevant to the modern industry.

Our fundamental position is that development of the Marcellus Shale need not produce winners and losers. If done right, the industry, the people of Pennsylvania, and the environment can all benefit from the combined effects of government regulation that is equal to the task at hand and the enforcement of best management practices in an industry that has already demonstrated its capabilities in this regard. It is our hope and belief that Pennsylvania can be a model to the nation in sustainable energy development, and preserve the historic landscape of Pennsylvania's environment for future generations.

IMPACTS OF DEVELOPMENT IN PENNSYLVANIA

Over the past five years, the development of the Marcellus Shale gas formation in Pennsylvania has increased at an ever-quickening pace. It has transformed the stagnant natural gas industry in Pennsylvania and has significant implications in the energy market at both the national and perhaps the international scale. It has also affected local communities, particularly communities not accustomed to industrial scale activity, in both positive and negative ways.

Last year 3,314 permits were issued by the Department of Environmental Protection, and 1,446 wells were drilled for the Marcellus Shale formation. Permit approvals for 2011 are already outpacing last year's numbers. Some analysts estimate that over the next 10 years annual Marcellus Shale formation well drilling activity will steadily increase from 2,500 wells per year to over 3,500 wells per year.[2]

The projected development of the Marcellus Shale formation and the anticipated development of the Utica Shale and perhaps other shale formations in Pennsylvania means that we are in the formative years of an industry that will be a prominent part of the Pennsylvania landscape for multiple generations to come.

One of the challenges of unconventional gas development in Pennsylvania is the density of well development activities. Thus far, well development activities have been concentrated primarily in a handful of counties in the northern tier and in the southwest corner of the Commonwealth. With very limited exception,[3] this concentrated activity is occurring without gathering information that is needed to identify and assess the impacts of this activity. PEC believes that as the natural gas exploration, production and delivery infrastructure is developed in the rest of the Marcellus Shale play and in other shale gas formations in Pennsylvania, it is vitally important to systematically collect and compile a data base of relevant information that is publicly accessible. Such information is necessary to assess the impacts of the past activity as well as ongoing activity, and to establish adaptable requirements that are designed to mitigate impacts to the greatest extent possible both in the near term and the long term.

There are a number of studies being undertaken by organizations such as the Environmental Protection Agency, the Department of Energy National Energy Technology Laboratory and the Interstate Oil & Gas Compact Commission to identify impacts of shale gas development. In addition, we are aware that:

- The U.S. Environmental Protection Agency is conducting a study plan on the potential environmental and human health implications of hydraulic fracturing, with special emphasis on the relationship between hydraulic fracturing and drinking water resources;
- President Obama has asked the Department of Energy's Advisory Board to produce a report offering recommendations on how to assure that shale gas development does not adversely affect water quality; and
- Last week the Chesapeake Bay Foundation petitioned the Council on Environmental Quality to conduct a multi-state Programmatic Environmental Impact Study examining the potential risks and possible cumulative impacts of natural gas development throughout the Marcellus Shale formation.

We are also aware that the state of New York has been engaged in a general environmental impact statement process relating specifically to unconventional wells developed by high volume hydraulic fracturing.

PEC's focus since the publication of our report has been the design of a model state-level regulatory and management process that operates on proactive, comprehensive information gathering and assessment prior to individual site development and well operation. In short, information gathering and assessment should be an integral part of the regulatory and permitting process. This basis then allows for adaptive management as greater understanding of this complex and dispersed activity – and its impacts – is developed.

REGULATORY CHALLENGES IN PENNSYLVANIA

Pennsylvania's regulatory framework as it existed five years ago was designed for shallow vertical wells that were far less complicated than horizontal shale gas wells.

The production of natural gas from what are termed "unconventional" resources such as the Marcellus Shale formation is a much more complex set of operations. The nature of these operations, which include the use of high volume hydraulic fracturing, as well as greater associated infrastructure, equipment and transportation demands, significantly increase the potential for adverse impacts to terrestrial and aquatic resources. What's more, any industrial process is subject to failures of technology and human judgment.

Given the rapidly increasing deployment of Marcellus Shale activity, often occurring in either close proximity to communities or sensitive natural resources or both, the need to reform Pennsylvania's management program has been great. In response, the Department of Environmental Protection has effected critical changes to its oil and gas management program and successfully proposed new regulations for promulgation by the Pennsylvania Environmental Quality Board.

On the regulatory front, DEP formulated and directed two major regulations through the Environmental Quality Board rulemaking process in the past two years:

- Regulatory amendments made to 25 Pa. Code Chapter 78[4] that significantly strengthen well casing and cementing standards to better ensure well integrity and protect against the migration of methane and hydraulic fracturing fluids. Insufficient well casing and completion has already proven to cause adverse impacts to private and public water supplies in Pennsylvania.
- Regulatory amendments made to 25 Pa. Code Chapter 95[5] that place stringent total dissolved solids (TDS) limits for the disposal of wastewater generated through unconventional shale gas development. These amendments have created strong economic incentive for the reuse of flowback fluids, helping to offset a still-significant water management challenge for Pennsylvania.

These vital changes will greatly contribute to better management of the industry and the avoidance of detrimental impacts.

But there remains much to be done, and we must acknowledge that the implementation of best management practices within the industry – beyond the point of regulation – are equally important to ensuring that unconventional shale gas development can provide economic growth and energy production without undue and long-term costs to the environment and people of Pennsylvania. At a minimum, this work must include:

- Ensuring that sufficient financial assurance is in place to address impacts if and when they arise. Currently Pennsylvania's bonding program is woefully inadequate in relation to the size and extent of unconventional shale gas operations.
- Enacting additional amendments to Pennsylvania's Oil & Gas Act to provide the Department of Environmental Protection with more precise authority to protect public and natural resources in the permitting and enforcement process.
- Enhancing Pennsylvania's Natural Heritage Program to better accommodate more regional or comprehensive planning by state resource agencies and the industry; as opposed to segmented, site-by-site analysis.
- Ensuring that sufficient funding is consistently provided to Pennsylvania's resource protection agencies so they may adequately perform their statutorily-mandated responsibilities. During the last several years, our state resource agencies have experienced the greatest burden of budget cuts and staffing reductions.[6]
- Ensuring better and more consistent management of water use in unconventional shale gas operations across Pennsylvania, including more comprehensive accounting and analysis of potential aquatic resource impacts.
- A restructuring of the overall permitting process which provides more robust and effective informational collection and assessment prior to the commencement of individual well operations.

We sincerely believe these objectives can be accomplished in a manner that is equally protective of the environment and public as it is supportive of the industry. We would like to address the latter two items in a bit more detail to highlight the nature of this challenge as well as how the objectives of PEC's work can better inform management efforts.

WATER MANAGEMENT

Current estimates place water demands for hydraulic fracturing at three to five million gallons per well. While the demands of unconventional shale gas development may be less than those of other industrial or energy producing activities, the fact remains that unconventional shale gas development is more likely to occur in remote locations where water withdrawals from smaller or high quality streams can quickly have significant cumulative impacts. Thus, the need for consistent and effective management across Pennsylvania is critical.

Yet management of water use for Marcellus Shale operations in Pennsylvania is a matter of geography. If a proposed water withdrawal is made in the Susquehanna or Delaware River Basins, it will be subject to the respective regulatory programs of the Susquehanna or Delaware River Basin Commissions. While the Susquehanna River Basin Commission has a robust water management program for Marcellus Shale operations, the Delaware River Basin Commission has imposed a moratorium while it works to update its regulations in response to growth of the industry.

In the Ohio River Basin, which does not have a corresponding Commission in place with water quantity authority, the Department of Environmental Protection has developed a Water Management Plan component as part of the well operation permitting process. The Department requires this submission through extension of existing authority via the Pennsylvania Clean Streams Law.[7]

The division of water management responsibility is unavoidable given that the River Basin Commissions are rightly acting pursuant to interstate compacts. The challenge is now before the Department and River Basin Commissions to work together to determine how water quantity issues can be best addressed, and to establish exemplary and consistent protocols and best management practices throughout Pennsylvania.[8]

PERMITTING

As indicated previously, Pennsylvania's regulatory framework for natural gas exploration and production, as it existed at the onset of the Marcellus Shale boom in 2005, did not contemplate the scale and intensity of horizontal unconventional well development activities. It was designed for conventional reservoir formation vertical wells which required limited, if any stimulation by hydraulic fracturing.

The Department of Environmental Protection has responded to some of the identified deficiencies in the conventional permitting framework through administrative alterations to the application process; the development of proposed amendments to oil and gas regulations and water quality standards which were promulgated by the Pennsylvania Environmental Quality Board; and other actions to increase the capacity to review applications and monitor well development activities. However, PEC believes that a more fundamental change in the permitting process is required because the existing process does not provide for the acquisition of sufficient information to make well-informed well pad siting decisions.

Based on discussions with unconventional gas exploration and development companies, we have the sense that some of the companies voluntarily implement sophisticated well pad siting processes that go beyond the minimum requirements in the current application process, involve meaningful engagement with surface property owners and the community, and are intended to mitigate potential impacts to the greatest extent reasonably possible. However, we also have the sense that the attitude and effort level of the sector as a whole varies widely.

Later this month PEC will present a package of proposed amendments to the Pennsylvania Oil and Gas Act, which will include a section that will fundamentally alter the existing permit application process. The revised application process is designed to gather more information on site conditions and focus more attention on the siting of well pads and associated infrastructure. Generally speaking, we will advocate that the permit application process for unconventional wells involving development by high volume hydraulic fracturing be split into two distinct phases. The first phase will be limited to the identification and assessment of site conditions for the purpose of determining whether a well pad should be authorized and, if so, the siting conditions that must be taken into account for selecting the precise location of the well pad and ancillary infrastructure. The second phase will focus on construction authorization of the well pad and the drilling, casing and development of the wells.

We believe that the two phase approach will allow for more flexibility in the siting of well pads to minimize risk and reduce the surface impacts of well pads and ancillary facilities to the greatest extent possible. At the same time, because the siting of the well pad will be pre-approved in the Phase I process, the Phase II application process should be at least as time-efficient, if not more efficient, than the current process. Consequently, once a well developer has assembled a portfolio of approved sites through the Phase I authorization process, it should have the necessary flexibility in planning rig movement.

In developing our proposal, we are cognizant of the interest that the process be reasonably predictable in terms of the level of effort required by the well developer to complete applications and the amount of time required to process pending applications. We believe that the two-phase process can be structured and implemented in a manner that will not unduly interfere with efficient well development.

BEYOND PENNSYLVANIA

As you can see, the breadth of issues in Pennsylvania is daunting, and our state is only one of many experiencing shale gas development. While we have found lessons learned and actions taken in other states to be instructive to our own work on these issues, they are tempered by the fact that Pennsylvania's topography, geology and climate are very different from places like Texas, Alabama, and Colorado. For this reason we believe that efforts to improve management and oversight of the industry should be primarily directed at the state level.

But that obviously does not and should not preclude the federal government from continuing its strong oversight of any shale gas state, including Pennsylvania. In fact, for some of the same reasons addressed before in this testimony – including the very real struggles of state agencies to meet existing state and federal mandates due to budgetary constraints – we believe federal engagement is essential.

Consistent with our principle of adaptive management, as information develops and if better understanding of individual and cumulative impacts point toward the need for revision of federal statues or regulations, we would urge swift and appropriate action. As Pennsylvania has learned first hand, environmental legacies from improperly controlled resource development is extraordinarily costly and detrimental to economic vitality and public well being. Even today Pennsylvania faces abandoned mine remediation costs that well exceed one billion dollars, and must account for thousands of miles of rivers and streams which fail to meet water quality standards because of acid mine drainage.

With the ongoing growth of the unconventional shale gas activity, we don't have a moment to lose in getting it right.

CONCLUSION

The oil and gas regulatory structure in Pennsylvania did not contemplate horizontal drilling in combination with high volume hydraulic fracturing and is not adequate to manage the escalating development of Marcellus Shale development throughout Pennsylvania. The current regulations are not designed to obtain timely and sufficient information to make well-informed decisions concerning the siting of well pads or to build a database identifying cumulative impacts of well development activities on the scale projected by the oil and gas industry. The natural gas industry has made great strides in leading innovation, but the regulatory framework must address the complexities created by the pressure of time, scale, cost and technology.

Given the extraordinary opportunities and challenges associated with Marcellus Shale gas extraction, it is incumbent upon key stakeholders to take whatever steps are necessary to ensure the safe and reliable development of this resource in a way that does not repeat the mistakes of the past.

Throughout Pennsylvania's history, our natural resources have been exploited for industrial purposes without the benefit of careful consideration and forethought. The price paid in exchange for this rapaciousness can never be fully calculated, yet remains evident in the forests, waterways, and communities and that cost has been shouldered by generations that followed the development.

More recently, accidents at drilling rigs have captured the attention of the news media, regulators and Pennsylvania citizens. These incidents cannot and should not be ignored – they highlight the need for prompt and effective reform.

The spot market for natural gas is considered to be temporarily undervalued, with gas prices of approximately $4.25 per million cubic feet. Even at that low price, the Marcellus Shale represents a natural resource whose development can be valued at $1- 2 billion in Pennsylvania. Considered in this context, PEC urges that a long-term view of development be adopted which allows all stakeholders to realize the benefits of the resource while safeguarding the health and safety of our citizens and the environment that has still not yet fully recovered from past resource development movements.

Pennsylvania has an extraordinary opportunity to enact the nation's best body of laws governing the extraction of a vast natural resource. Such action would effectively legislate the nation's best practices and make them the standard by which the Marcellus Shale is developed and provides the benefits to the Commonwealth that have been heralded as the promise of this new industry.

End Notes

[1] Available at http://www.pecpa.org/marcellus

[2] See Timothy J. Considine, et al, The Economics of the Pennsylvania Marcellus Shale Natural Gas Play: An Update, at 16, (The Pennsylvania State University, College of Earth and Mineral Sciences, Department of Energy and Mineral Engineering, May 24, 2010).

[3] In November 2010 the Pennsylvania Chapter of The Nature Conservancy published its first in a series of analyses on energy development impacts to Pennsylvania. This first report included an analysis of potential impacts from Marcellus Shale development. The report is available at: http://www.nature.org/media

[4] Pennsylvania Bulletin August 21, 2010 (40 Pa.B. 4835)

[5] Pennsylvania Bulletin February 5, 2011 (41 Pa.B. 805)

[6] One notable exception has been the Department of Environmental Protection's concerted effort in adding Oil and Gas Bureau field staff to monitor Marcellus Shale activity over the past several years. However, imposed budget cuts continue to significantly affect other Department Bureaus and Programs, many of which have a role to play in overall management of this escalating activity.

[7] Pa. Stat. Ann. Tit. 35, §691.1 et seq. The Clean Streams Law does not directly provide for regulation of water withdrawals. Rather its focus is on activities that cause or may cause "pollution" (broadly defined to include physical, chemical or biological alteration) to waters of the Commonwealth. *See generally* R. Timothy Weston, *Water and Wastewater Issues*, Prepared for the 2011 Penn State Marcellus Shale Law and Policy Symposium (February 10, 2011). One of the recommendations of PEC's report is to provide clear statutory authority for the Department to manage large scale water withdrawals.

[8] The program established by the Susquehanna River Basin Commission is frequently cited as a model for how effective management with strong informational reporting can be performed without unduly affecting the industry.

Chapter 9

TESTIMONY OF ROBERT M. SUMMERS, PHD., ACTING SECRETARY OF THE MARYLAND DEPARTMENT OF THE ENVIRONMENT, BEFORE THE SUBCOMMITTEE ON WATER AND WILDLIFE, HEARING ON "HYDRAULIC FRACTURING IN THE MARCELLUS SHALE AND WATER QUALITY"[*]

Chairman Boxer, Chairman Cardin, Ranking members Inhofe and Sessions and honorable members of the Committee, thank you for the opportunity to share Maryland's experience and concerns with hydraulic fracturing in the Marcellus Shale.

THE MARCELLUS SHALE IN MARYLAND

The Marcellus Shale formation underlies Garrett County and part of Allegany County in the far western portion of Maryland. In these two counties,

---

[*] This is an edited, reformatted and augmented version of testimony given by Robert M. Summers, PhD., Acting Secretary of the Maryland Department of the Environment, before the Subcommittee on Water and Wildlife, Hearing on "Hydraulic Fracturing in the Marcellus Shale and Water Quality" on Tuesday, April 12, 2011.

gas companies have leased the gas rights on more than 100,000 acres. The Maryland Department of the Environment issues permits for oil and gas wells, and we received our first permit application for drilling and hydraulic fracturing ("fracking") in the Marcellus Shale in 2009. No permits have yet been issued. We currently have applications pending from two companies for a total of 5 wells. We are mindful of the tremendous benefits that could accrue to the environment and the economy by exploring and exploiting our gas reserves, but we are equally alert to the risks of adverse public health and environmental effects. Our paramount concern is protecting our ground and surface waters.

Having observed events in Pennsylvania during the first few years of Marcellus Shale drilling there, Governor O'Malley, the Department of the Environment, and the Department of Natural Resources are determined to ensure that drilling will not start in Maryland until we know whether, and how, it can be done safely. We are proceeding cautiously and deliberately and do not intend to allow drilling and fracking in the Marcellus Shale until the issues are resolved to our satisfaction.

An industry representative estimated that as many as 1,600 wells could be drilled in 128,000 acres in Garrett County and 637 wells in 51,000 drillable acres in Allegany County. There is a huge potential economic impact. Lease payments, royalties, and in Garrett County, severance taxes, and the economic activity associated with drilling-related jobs could bring an economic boom to these western counties and some of their citizens. The consequences of a later economic collapse and the cost of the potential environmental damage are harder to quantify.

Although Maryland has not permitted any Marcellus wells, the Department of the Environment has been attentive to the possible shipment of fracking fluid into Maryland since late 2008. Some flow back from fracking in another state was shipped to Baltimore for treatment and disposal in 2009. The fracking fluid was pretreated and sent to a large municipal wastewater treatment plant that discharges to brackish water and not upstream of any drinking water intake. For these reasons, and because of the small volume of fracking fluid relative to the flow from the wastewater treatment plant, this handling posed little or no risk. A different situation could exist, however, if concentrated fracking water were not treated adequately and discharged upstream of a drinking water intake. Concurrently, we have had discussions with EPA Region III, which is advising states on monitoring to ensure that drinking water remains safe.

ENVIRONMENTAL, PUBLIC HEALTH AND PUBLIC SAFETY CONCERNS

There are numerous issues that need to be addressed before Maryland can conclude whether and how drilling in the Marcellus Shale can be done safely. They include:

- minimum requirements for constructing, casing and cementing wells
- minimum requirements for integrity testing of wells
- minimum requirements for installing and testing blowout prevention equipment
- the potential migration of gas from the well, including migration from induced or naturally occurring faults and fractures
- the toxicity, fate and transport of fracking fluid
- proper handling and disposal of naturally occurring radioactive materials
- best practices for managing and disposing of flow back
- best practices for managing and disposing of drilling mud and drill cuttings
- best practices for containment and management of fuels and other liquids
- air pollution, including ozone production
- re-fracturing and its potential effect on well integrity
- avoiding habitat fragmentation, invasive species, and damage to wetlands and streams from access roads, drill pads, gathering lines, and ancillary operations
- avoiding other impacts to aquatic ecosystems, including stream sedimentation from damaged roads and dust from truck traffic
- the adequacy and sustainability of surface water and ground water in the region to supply water for fracking
- public safety and emergency response services

MARYLAND LEGISLATION

Public concern brought the issue of Marcellus Shale drilling to the attention of Maryland legislature, which started its 90-day session on January 12, 2011. Bills were introduced to speed the issuance of drilling permits, place the burden on each applicant for a permit to demonstrate the safety of drilling and fracking, and require a study before permits could be issued. The Governor and the Department supported a bill to require the State to perform a comprehensive study of short-term, long-term and cumulative effects of hydraulic fracturing, to be paid for by those gas companies holding leases in Maryland. Until publication of the report, the legislation would prohibit the Department from issuing a permit involving hydraulic fracturing unless it can be done without adverse impact to human health, natural resources, or the environment. As this is being written, the fate of these various bills in the Maryland legislature in unknown.

HOW THE MARYLAND DEPARTMENT OF THE ENVIRONMENT PROPOSES TO PROCEED

We anticipate moving forward in two stages. First, during the next year, we will survey existing practices and select Best Practices for the drilling and fracking of wells. These Best Practices will cover all aspects of site preparation and design, delivery and management of materials, drilling, casing, cementing and fracking. After we develop this interim "gold standard" the Department will consider issuing permits for a small number of exploratory wells to be drilled and fracked in the Marcellus Shale using these standards. Sites eligible for these exploratory permits must present minimum risks to human health and the environment. The permit will be conditioned on the company's commitment to collect and share with the State data from drilling, fracking and monitoring to advance our understanding of the risks and the adequacy of the Best Practices.

Second, we will use the data from these exploratory wells, along with the results of other research as it becomes available, to evaluate the environmental viability of gas production from the Marcellus Shale. This phase will focus on long-term and cumulative risks, and include landscape level effects like forest fragmentation. If we determine that gas production can be accomplished without unreasonable risk to human health and the environment the Department could then make decisions on applications for production wells. Permit conditions would be drafted to reflect Best Practices and avoid environmental harm. At this time, the State has not identified a source of funding for this work, other than the proposed legislation mentioned above.

THE NEED FOR FEDERAL LEADERSHIP

We need the federal government to take a more active role in studying and regulating activities such as deep drilling, horizontal drilling, hydraulic fracturing, and waste disposal. While the states should retain the authority to enact more stringent requirements, a federal regulatory "floor" would ensure at least basic protection of the environment and public health. In previous administrations, the balance has been struck in favor of energy production over environmental protection. For example, gas and oil exploration and production wastes are excluded from RCRA Subtitle C regulation. The injection of hydraulic fracturing fluids is excluded from the Safe Drinking Water Act's Underground Injection Program. The Clean Water Act was amended to expand the exemption of stormwater runoff to cover all oil and gas field activities and operations, not just uncontaminated stormwater runoff from certain operations. In the absence of a strong federal regulatory program, the burden of assuring that wells can be safely drilled and hydraulically fractured in the Marcellus Shale falls on the states individually. Maryland believes that federal technical support and oversight of state regulatory programs such as those administered under the Clean Water Act and the Safe Drinking Water Act are particularly important to ensure appropriate protection of interstate waters such as the Susquehanna and Potomac Rivers and the Chesapeake Bay, which are critical resources to all of the jurisdictions in the region.

We commend Congress for directing the United States Environmental Protection Agency (EPA) to conduct research to examine the relationship between hydraulic fracturing and drinking water resources. EPA's Office of Research and Development has developed a solid, comprehensive plan for this study; however, we note that some important issues are beyond the scope of the study, including re-fracturing, and impacts to air quality and terrestrial and aquatic ecosystems.

We are also encouraged by President Obama's "Blueprint for a Secure Energy Future," which he announced on March 30. In particular, we welcome the plan to have the Energy Advisory Board establish a subcommittee to identify immediate steps that can be taken to improve the safety and environmental performance of fracking and to develop consensus recommendations for federal agencies on practices that will ensure the protection of public health and the environment. The offer of technical assistance from DOE and EPA is also welcome.

The states need the federal government to lead and to lend its resources to the effort and we need a strong state-federal partnership. Timing and other factors probably preclude using an exploratory well in Maryland for one of the prospective case studies planned for the EPA report, but we intend to seek EPA guidance on the study plan for the prospective case study so that Maryland can gather the most relevant data if a permit is issued for an exploratory well. We also intend to seek technical assistance from the USGS in determining what to monitor in the process of drilling and fracking wells for exploration, and in analyzing the data we obtain. Preliminary guidance from EPA on the proper spatial area for monitoring and recommendations for Best Practices to prevent environmental impacts from drilling and fracking operations would be very helpful until the EPA study can be completed. Lastly, EPA should develop water quality criteria for conductivity (specific to chemical species), dissolved solids and salinity in freshwater, as well as pretreatment standards and effluent limitations for fracking flowback.

Under existing federal law, hydraulic fracturing is excluded from Safe Drinking Water Act regulation of underground injection. The chemicals added to fracking fluid do not have to be disclosed. We support the Fracturing Responsibility and Awareness of Chemicals Act, S.587, which was introduced on March 15, 2011, by Senator Casey and co-sponsored by Senator Cardin. The Bill would reinstate regulation of hydraulic fracturing under the Safe Drinking Water Act and require the person conducting hydraulic fracturing operations to disclose to the government all of the chemical constituents used in hydraulic fracturing. Proprietary chemical formulas could still be protected from public disclosure. These are positive steps, although we encourage a reexamination of scope of protection for proprietary information. The public has an important interest in knowing what chemicals are being injected underground.

The Chesapeake Bay Foundation and other groups have filed a petition with the federal government for a Programmatic Environmental Impact Statement to address the risks and cumulative impacts of the extraction of natural gas from the Marcellus Shale formation in the Chesapeake Bay watershed. We support the goal of a comprehensive assessment, and we note that portions of the Marcellus Shale lie to the west of the Eastern Continental Divide, and that the environment outside the Chesapeake Bay watershed deserves protection, too.

Thank you for taking the initiative to inquire into this important issue and for providing the opportunity to share Maryland's perspective.

Chapter 10

TESTIMONY OF THE HONORABLE JEFF CLOUD, OKLAHOMA CORPORATION COMMISSION, VICE CHAIRMAN, BEFORE THE SUBCOMMITTEE ON WATER AND WILDLIFE, HEARING ON "NATURAL GAS DRILLING: PUBLIC HEALTH AND ENVIRONMENTAL IMPACTS"[*]

I very much appreciate the opportunity to testify today before the joint hearing of the Senate Committee on Environment and Public Works and the Subcommittee on Water and Wildlife about the regulation of hydraulic fracturing and Oklahoma's many decades of experience in this regard.

Oklahoma's first commercial oil well was drilled in 1897, which was 10 years before Oklahoma officially became a state in 1907. Since then, oil and natural gas production has expanded into almost part of the state.

The Oklahoma Corporation Commission (OCC) was first given responsibility for regulation of oil and gas production in Oklahoma in 1914. Currently the Commission has exclusive state jurisdiction over all oil and gas industry activity in Oklahoma, including oversight and enforcement of rules

---

[*] This is an edited, reformatted and augmented version of testimony given by Honorable Jeff Cloud, Oklahoma Corporation Commission, Vice Chairman, before the Subcommittee on Water and Wildlife, Hearing on "Natural Gas Drilling: Public Health and Environmental Impacts" on TUESDAY, APRIL 12, 2011.

aimed at pollution prevention and abatement and protecting the state's precious water supplies.

Presently, there are over 185,000 active wells in Oklahoma – roughly 115,000 oil, 65,000 gas and 10,500 injection/disposal wells – and thousands of miles of gathering and transmission pipelines.

In recent years the Woodford Shale in Oklahoma has become an important source of natural gas for the nation. The development of the Oklahoma's Woodford Shale, like the other shale regions in the United States, has been made possible by horizontal drilling and hydraulic fracturing technologies.

Hydraulic fracturing has been used for over 60 years in Oklahoma, and more than 100,000 Oklahoma wells have been hydraulically fractured over that period. Over that more than half century of hydraulic frac experience, there has not been a single documented instance of contamination to groundwater or drinking water as result of hydraulic fracturing.

To say we take protection of our water quality seriously would be an understatement. Our rules are constantly reviewed and updated with that in mind. Our rules include a general prohibition against pollution of any surface or subsurface fresh water from well completion activities.

Proper casing and cementing represent the primary means of protecting fresh water during hydraulic fracturing operations. Production casing is required to be cemented a minimum of 200 feet above oil and gas producing zones and Corporation Commission inspectors must be given at least 24 hours advance notice of cementing so that the inspector can witness that activity.

Oklahoma Corporation Commission rules address procedures in the event of unanticipated operational or mechanical changes that may be encountered during hydraulic fracturing and require the operator of the well to contact Commission officials within 24 hours of discovery of a casing problem. In addition to that notice, immediate remedial action is also required to repair any problems with surface or production casing.

A guidance document, referred to as the Oklahoma Corporation Commission "Guardian Guidance" has been developed and distributed widely. It provides the incremental process that an oil and gas operator should follow to assess, remediate if necessary and to close any site that would be found to impact ground water or surface water.

As previously stated, Corporation Commission rules require operators to give 24-hour notice before setting surface casing or cementing surface strings. Standard Commission rules also require an operator to submit a well completion report within 30 days after completion activities. The volumes of fluids and proppants used in any hydraulic frac operation are required on the form. Commission rules also allow the agency to obtain information identifying chemicals used in hydraulic fracturing or other exploration and production operations.

In August of 2010, the Oklahoma Corporation Commission volunteered to have its hydraulic fracturing program reviewed by a 12-year-old multi-stakeholder organization known as STRONGER, or, by its full name, State Review of Oil & Natural Gas Environmental Regulations. The Oklahoma oil and gas regulatory program had undergone three successful prior reviews.

From October 2010 through January 2011, a sevenperson, again multi-stakeholder review team appointed by STRONGER conducted an in-depth examination of Oklahoma's hydraulic regulatory program. The review team included Leslie Savage of the Texas Railroad Commission; Wilma Subra, of Subra Company of New Iberia, Louisiana (and a noted critic of the domestic oil and gas industry); and Jim Collins of the Independent Petroleum Association of America. The official observers included the Oklahoma Sierra Club and the United States Environmental Protection Agency Region VI.

The review team concluded that the Oklahoma program is, overall, well managed, professional and meets its program objectives. Incidentally, the U.S. Environmental Protection Agency and U.S. Department of Energy have provided grant funding to STRONGER to support its activities.

I would also note that in Oklahoma, collaboration involving the regulated oil and gas industry, other stakeholders and my state agency's Staff have repeatedly led to successful development of rules and policies to address environmental protection issues, particularly the protection of water.

An example: In 2008 when the Corporation Commission modified and expanded rules over oil and gas industry activity to control runoff of soil and dissolved minerals and chlorides into the watershed that feeds two vital southeast Oklahoma lakes.

The two particular lakes that I mentioned – Lake Atoka and McGee Creek Reservoir – are exceptionally clean and provide very high quality water to the City of Oklahoma City about 100 miles away. But the lakes also are on top of deep rock deposits that hold huge amounts of natural gas, which, in the best interest of Oklahoma and the nation, we want to allow the petroleum industry to find and produce.

The rules established in 2008 replaced out-dated 1985 field rules that were developed back when there was much less oil and gas industry activity in that part of the state. The new rules were hammered out over several meetings involving representatives of Oklahoma City's water utility division, the oil and gas industry, rural water districts, counties, tribes, my agency's regulatory enforcement Staff and other stakeholders.

These revised rules modified requirements for setbacks (buffer zones) from lakes, retaining walls or berms around wellsites, for pit liners and operations for muds and fluids used and produced during drilling operations and for other requirements to provide flexibility needed by industry while increasing protection of water resources.

This allows the oil and gas industry to continue to pursue its important goal of finding and producing critical natural energy resources while also providing added measures to ensure the quality of lake waters and environment are protected.

Without the need for any federal intervention, the City of Oklahoma City, the regulated oil and gas industry and the State worked together to come up with acceptable protections because we all realize that it is in our mutual best interest to ensure proper and practical water and environmental protections without cutting off access to critical resources.

Nature by itself unfortunately did not bless Oklahoma with any natural large bodies of water, so fresh water is especially precious in my state. Oklahoma has more than 50 man-made lakes. It is worth noting that Texas is currently in court suing Oklahoma to get our state's water. We must be doing something right.

All of us can agree that there needs to be "rules of the road" and that those rules need to be followed and enforced. The issue is what works best in making sure that those rules are followed and that Oklahoma's water and our environment are protected. Oklahoma's record makes it clear that state regulation is the best way to meet those goals I and my two fellow Commissioners hold elected positions. We are directly accountable to our fellow Oklahomans. We have both a vested and personal interest in ensuring our water is protected. After all, and not to be trite, we drink the water, too.

In: Hydraulic Fracturing and Natural Gas Drilling   ISBN: 978-1-61470-180-4
Editor: Aarik Schultz   © 2012 Nova Science Publishers, Inc.

Chapter 11

TESTIMONY OF DAVID NESLIN, DIRECTOR, COLORADO OIL AND GAS CONSERVATION COMMISSION, BEFORE THE SUBCOMMITTEE ON WATER AND WILDLIFE, HEARING ON "NATURAL GAS DRILLING: PUBLIC HEALTH AND ENVIRONMENTAL IMPACTS"[*]

Madame Chair, thank you for this opportunity to provide our perspective on how the state of Colorado is protecting public health and the environment while we develop our important, indigenous oil and gas resources. My name is David Neslin, and I am the director of the Colorado Oil and Gas Conservation Commission, the state agency responsible for regulating oil and gas development. I thank you and your colleagues for your thoughtful consideration and your efforts to gather the information necessary to properly evaluate these matters.

Colorado has a long and proud history of oil and gas development, with our first oil well dating back to 1862. As of 2009, we ranked fifth in natural gas production and tenth in oil production. Our diverse hydrocarbon resources encompass a variety of shale, tight sand, coal bed methane, and other formations that span the state. At the same time, we have a thriving resort and

---

[*] This is an edited, reformatted and augmented version of testimony given by David Neslin, Director, Colorado Oil and Gas Conservation Commission, before the Subcommittee on Water and Wildlife, Hearing on "Natural Gas Drilling: Public Health and Environmental Impacts" on Tuesday, April 12, 2011.

tourist economy, and our rugged mountains, clear streams, and abundant wildlife are an essential part of our heritage.

I want to focus most of my comments today on hydraulic fracturing, as it remains at the center of discussion on oil and gas drilling both in Colorado and nationally. Most of Colorado's 44,000 active oil and gas wells, as well as the thousands of new wells that will be drilled in the coming years, rely on hydraulic fracturing to create the permeability that allows fluid and gas to flow into the wellbore and be produced. It is not an understatement to say that this technology is absolutely vital to unlocking Colorado's rich natural gas and oil reserves. These reserves are a critical source of domestic energy for our state and nation, and their exploration, development, and production provides good-paying jobs for our residents and needed tax revenues for our communities.

But it is also essential that this development occurs in an environmentally responsible manner that protects our water resources generally and our drinking water specifically. This is a fundamental part of our regulatory mission, and something that everyone at our agency takes very seriously. To this end, our environmental professionals have investigated hundreds of groundwater complaints over the years. To date, we have found no verified instance of hydraulic fracturing harming groundwater. These investigations have been documented and are publicly available on our website.

In addition, since 2000, our Commission has required operators to collect pre- and post-development water quality samples from more than 1,900 water wells in the San Juan Basin in Southwestern Colorado, which has historically been one of our most productive natural gas-producing areas. Thousands of oil and gas wells in that Basin have been hydraulically fractured, and if fracturing fluids were reaching these water wells then you would expect changes in the chemical composition of the water. However, independent analysis of the data has found no statistically significant increase in chemical concentrations. A report documenting this analysis is likewise available on our website, and we have offered to share this data with the Environmental Protection Agency. We have also collected or required operators to collect similar data from about another 1,900 water wells in other gas and oil producing areas of the state.

I would also like to emphasize that during 2007 and 2008, our agency devoted substantial time and effort to updating our regulations to address a broad range of environmental issues associated with oil and gas development. This rulemaking process lasted 16 months, included testimony from 160 witnesses, and involved 22 days of hearings. The final rules strike a responsible balance between energy development and environmental protection, and they reflect input from dozens of local governments, oil and gas companies, and environmental groups, as well as thousands of our residents. And many other states have subsequently taken or are taking similar action, including Wyoming, Oklahoma, Ohio, Pennsylvania, and Arkansas.

These recent state rulemakings exemplify the benefits associated with state oversight and site-specific regulation, and they have specifically addressed hydraulic fracturing concerns. Colorado's amended rules contain various provisions to ensure that such activities do not harm our drinking water while recognizing the variety of issues at play in different regions of our state. For example,

- Rule 205 requires operators to inventory chemicals kept at drilling sites, including hydraulic fracturing fluids. This information must be provided to agency officials promptly upon request and also to certain health care professionals who sign a confidentiality agreement. This allows government officials and medical professionals to investigate and address allegations of chemical contamination associated with hydraulic fracturing, while protecting proprietary information.
- Rule 317 requires wells to be cased with steel pipe and the casing to be surrounded by cement to create a hydraulic seal and ensure that gas and fluids do not leak into shallower aquifers. Further, operators are required to run cement bond logs on all production casing to confirm that the cement has properly isolated the hydrocarbon bearing zones. Rule 341 requires operators to monitor well pressures during hydraulic fracturing and promptly report significant increases. Together, these requirements help to ensure that ground water is protected and that prompt action is taken if conditions arise that could lead to the subsurface release of hydraulic fracturing fluids.
- Rule 317B imposes mandatory setbacks and enhanced environmental protections on oil and gas development occurring near sources of public drinking water. These requirements provide an extra layer of protection for our public water supplies and help ensure that these critical resources are not inadvertently contaminated by energy development.
- Rule 608 requires operators developing coalbed methane wells to pressure test the wells and to sample nearby water wells before, during, and after operations to ensure that they are not contaminated by gas or other pollutants. Rule 318A requires operators in the DJ Basin in northeastern Colorado to do similar water well sampling

before drilling. These rules provide an extra layer of protection for water wells located near oil and gas development.
- Rules 903, 904, and 906 impose updated requirements for pit permitting, lining, monitoring, and secondary containment to ensure that fluids in pits do not contaminate soil, groundwater, or surface water. These requirements will help ensure that any flowback of hydraulic fracturing fluids is properly contained.

These regulations are important, and we believe they have substantially improved our protection of water resources. But we have not stopped there. We are continuing to take proactive, cost-effective steps to ensure oil and gas development, and hydraulic fracturing in particular, protects public health and the environment. Let me address these here.

First, we and other states have worked closely with the Groundwater Protection Council and the Interstate Oil and Gas Compact Commission on the launch of a new website – www.fracfocus.org. This site, launching this week, encourages oil and gas operators to voluntarily provide information on the chemicals they use to hydraulically fracture a well. As you know, this has been a sensitive issue for the public and the industry, and we believe this online chemical registry will provide helpful information to citizens who want to better understand hydraulic fracturing or have questions about a particular well. Under our regulations that took effect in 2009, operators already must disclose fracturing constituents upon request by state regulators or health professionals. The website will compliment this requirement by providing a wealth of information to the public, including company names, well locations, construction details, fracturing fluid constituents, and chemical abstract numbers.

Second, we have arranged to have our hydraulic fracturing regulations professionally audited this summer by STRONGER – The State Review of Oil & Natural Gas Environmental Regulations. STRONGER is a national organization consisting of state regulators and industry and environmental representatives. Their review process is a collaborative undertaking involving an evaluation of our regulations and a comparison of them against a set of guidelines developed and agreed to by all participating parties. During the last eight months, STRONGER completed similar reviews of the Oklahoma, Pennsylvania, Ohio and Louisiana hydraulic fracturing programs, and we are subjecting Colorado's program to a STRONGER review to determine whether further improvements can be made.

Third, the House of Representatives Committee on Energy and Commerce recently reported that about 1.3 million gallons of diesel fuel or fluids containing diesel fuel was used for hydraulic fracturing in Colorado during the period from 2005 through 2009. In response, we have launched our own investigation into this subject. This work involves both reviewing our own records as well as gathering information from service companies and operators. While we believe our regulations would have prevented contamination of drinking water supplies, we are collecting information independently to assess the situation.

Fourth, we continue to consider and assess public concerns that have arisen over hydraulic fracturing. As part of this, we have endeavored to give these matters the transparency they require and deserve. In February of this year, our Commission held a full public hearing to examine an allegation that hydraulic fracturing had contaminated a water well in southern Colorado. In this case, our Commissioners – a diverse board representing environmental, industry, local government and other sectors – unanimously determined that hydraulic fracturing had not impacted the well in question.

In summary, I want to stress how seriously we take this subject, and how Colorado is committed to ensuring that hydraulic fracturing protects public health and the environment. I also want to note that we are not unique, and that many other states are taking similar action. Our experience, and that of other states, demonstrates how hydraulic fracturing and other oil and gas activities are most effectively regulated at the state level, where highly diverse regional and local conditions are more fully understood and where rules can be tailored to fit the needs of local basins, environments and communities. In this way, we can ensure that our precious natural resources and environment are protected while providing our state and nation with a cleaner burning and vital source of domestic energy and allowing for the innovation and experimentation that are the hallmarks of our nation.

INDEX

A

abatement, xiv, 168
access, 5, 8, 62, 72, 73, 105, 118, 131, 162, 171
accounting, 40, 151
acetaldehyde, 71
acetic acid, 135
acid, 61, 65, 68, 74, 75, 76, 77, 79, 80, 81, 82, 83, 84, 85, 86, 87, 89, 90, 91, 92, 94, 95, 96, 97, 131, 154
acrylate, 81
additives, 25, 39, 53, 62, 99
Administrative Procedure Act, 23
adults, 72, 126, 128, 129
advancements, xii, 103
adverse effects, 130, 145
advocacy, 56
agencies, 34, 36, 38, 39, 40, 41, 43, 45, 48, 55, 56, 58, 108, 119, 139, 150, 154, 165
air emissions, 120
air pollutants, 61, 68, 71, 72, 73, 118, 128
air quality, xiii, 8, 34, 117, 118, 128, 164
Alaska, 19, 34, 114, 144
alcohols, 76, 83, 90
alkane, 88
amine, 75, 76, 78, 80, 81, 83, 84, 88, 93
amines, 76, 83, 84, 86, 93
ammonia, 85
ammonium, 76, 82, 83, 85, 87, 89, 90, 91, 93, 96

amorphous precipitate, 93
antimony, 140
appropriations, 2, 40, 43
aquatic life, 104
aquatic organisms, xiii, 125, 128
aquifers, 7, 8, 11, 13, 23, 55, 108, 118, 131, 132, 176
aromatic hydrocarbons, 140
assessment, 26, 33, 46, 56, 73, 148, 151, 153, 166
assets, 48
authorities, 121, 123, 135, 138, 141
authority, 2, 8, 11, 18, 19, 21, 25, 27, 45, 52, 56, 123, 138, 146, 150, 151, 156, 164
avoidance, 34, 150

B

bacteria, 48, 63
ban, 23, 108
barium, 132, 133, 134, 138
base, 86, 87, 147
bauxite, 46
benefits, xii, xiii, 9, 30, 32, 44, 45, 103, 119, 122, 124, 145, 155, 156, 160, 175
benzene, 53, 61, 69, 70, 99, 104, 118, 133, 135
bicarbonate, 77, 94
bioaccumulation, 133
blood, 67, 72

blood pressure, 72
bonding, 150
budget cuts, 110, 150, 156
by-products, 135

C

calcium, 82, 96, 97
cancer, 71, 127, 135
carbohydrates, 66
carbon, 15, 17, 41, 42, 57, 144
carcinogen, 61, 69, 70, 99, 110, 127, 135
case studies, 43, 165
case study, 21, 165
castor oil, 92
cellulose, 91
central nervous system, 70
ceramic, 46
ceramic materials, 46
CERCLA, 8, 48
challenges, 9, 42, 48, 118, 145, 147, 155
chemical, 29, 30, 43, 53, 54, 60, 61, 62, 64, 65, 66, 67, 70, 72, 73, 99, 120, 156, 165, 174, 175, 177
chemicals, xi, 1, 2, 3, 7, 24, 28, 29, 30, 37, 53, 54, 60, 61, 62, 63, 64, 65, 66, 67, 69, 70, 71, 72, 73, 99, 101, 105, 109, 110, 118, 127, 131, 133, 140, 165, 169, 175, 177
children, 71, 104, 126, 127, 128, 129, 134, 136, 138
chlorine, 95, 135
chronic drinking, 136
citizens, 12, 31, 105, 118, 126, 129, 155, 177
civil action, 12, 31
classes, 14, 15, 22, 133
Clean Air Act, 61, 68, 71, 100, 120, 122, 123
clean energy, 73
cleanup, 44, 47
climate, 124, 154
closure, 37
CMC, 134, 137

coal, xii, xiv, 1, 3, 4, 6, 23, 24, 25, 31, 32, 35, 42, 52, 54, 107, 121, 128, 144, 174
coalbed methane (CBM), 2, 4, 23
coffee, 61, 65, 87
collaboration, 170
color, iv
combined effect, 146
commercial, xiv, 54, 110, 127, 167
communities, 8, 24, 45, 118, 123, 137, 145, 146, 147, 149, 155, 174, 178
community, 119, 121, 153
competition, 24
competitors, 29
complement, 122
complexity, 9
compliance, 10, 24, 33, 45, 123
composition, 47, 62, 72, 120, 174
compounds, 7, 9, 53, 54, 61, 66, 70, 71, 73, 77, 90, 93, 99, 110, 131, 133
conductivity, 131, 165
conference, 145
confidentiality, 54, 175
Congress, 1, 2, 10, 19, 24, 27, 28, 34, 43, 45, 46, 49, 56, 57, 60, 63, 105, 119, 120, 164
consensus, 146, 165
conservation, 48, 55, 56
constituents, 26, 29, 39, 53, 71, 165, 177
construction, 9, 14, 15, 32, 37, 38, 54, 118, 131, 153, 177
consumers, 116
consumption, 13, 127, 134, 136, 140
containers, 132
contaminant, 23, 26, 49, 61, 137, 140
contaminated water, 7
contamination, 2, 3, 7, 8, 9, 24, 26, 30, 32, 34, 38, 39, 42, 43, 44, 48, 98, 105, 108, 109, 113, 114, 115, 118, 168, 175, 178
Continental, 166
copolymer, 75, 78, 84, 92
corrosion, 65, 136
cost, 26, 41, 55, 122, 155, 160, 176
Court of Appeals, 2, 20, 22, 27, 28, 54
covering, 9, 35
cracks, 21

criteria continuous concentration (CCC), 134, 137
criteria maximum concentration (CMC), 134, 137
critical infrastructure, 126, 129
crude oil, 122

D

danger, 48
database, 64, 155
DBP, 135
deep shale deposits, xiii, 125
defects, 71
deficiencies, 152
denial, 20
Department of Energy, 4, 47, 48, 55, 56, 58, 98, 147, 148, 156, 169
Department of Environmental Protection (DEP), 137
Department of Health and Human Services, 70, 100
deposits, xiii, 115, 125, 131, 143, 170
depth, 14, 53, 169
derivatives, 79, 89, 93, 96, 140
destruction, 67
diesel fuel, 2, 11, 25, 26, 27, 52, 63, 71, 110, 120, 121, 177
discharges, 121, 161
disclosure, 2, 28, 29, 30, 53, 60, 64, 99, 105, 166
disinfection, 135
displacement, 128
disposition, 131
District of Columbia, 28
domestic energy options, xiii, 117, 124
draft, 24, 26, 35, 58, 120
Draft Hydraulic Fracturing Study Plan, 43, 98
drainage, 132, 140, 141, 154
drilling fluids, 7, 105
drinking water, xi, 2, 3, 7, 8, 10, 11, 12, 13, 14, 18, 22, 23, 24, 25, 26, 31, 36, 43, 46, 48, 49, 50, 51, 54, 58, 60, 63, 67, 70, 71, 98, 104, 108, 109, 113, 118, 119, 121, 126, 127, 128, 133, 134, 135, 136, 137, 138, 139, 141, 148, 161, 164, 168, 174, 175, 176, 178

E

economic activity, 160
economic boom, 160
economic development, 5, 144
economic growth, 116, 150
effluent, 110, 121, 127, 128, 132, 133, 134, 135, 136, 137, 138, 139, 140, 141, 165
effort level, 153
emergency, 12, 54, 129, 162
emergency planning, 54
emergency response, 54, 129, 162
emission, 71, 122
employees, 65
employers, 65
energy, xii, xiii, 33, 44, 47, 50, 55, 56, 58, 62, 73, 98, 103, 105, 107, 115, 116, 117, 118, 120, 122, 123, 124, 143, 144, 146, 147, 150, 151, 156, 164, 170, 174, 175, 176, 178
Energy Information Administration (EIA), xii, 60, 98
Energy Information Agency (EIA), 4
Energy Policy Act of 2005, 2, 11, 120
energy prices, xiii, 44, 117
energy supply, 33, 122
enforcement, xiv, 9, 11, 12, 18, 19, 28, 32, 41, 45, 109, 110, 123, 139, 146, 150, 168, 170
Enhanced Oil Recovery, 17
environment, xiii, 49, 55, 60, 63, 64, 65, 73, 108, 117, 118, 123, 124, 130, 145, 146, 150, 151, 155, 160, 162, 163, 164, 165, 166, 171, 173, 177, 178
environmental impact, xi, 26, 35, 103, 104, 111, 118, 119, 123, 124, 148, 165
environmental issues, 8, 105, 119, 175
environmental organizations, 146
environmental policy, xiii, 125, 127
environmental protection, 119, 120, 164, 170, 171, 175, 176

Environmental Protection Agency, viii, 2, 4, 17, 18, 19, 47, 49, 50, 51, 52, 54, 57, 63, 98, 100, 117, 134, 147, 148, 164, 169, 175
environmental resources, xiii, 125
EPA, 2, 3, 4, 8, 9, 10, 11, 12, 13, 14, 15, 16, 17, 18, 19, 20, 21, 22, 23, 24, 25, 26, 27, 28, 29, 30, 31, 32, 33, 34, 37, 40, 41, 42, 43, 44, 45, 46, 48, 49, 50, 52, 54, 55, 57, 58, 63, 67, 70, 71, 98, 99, 100, 105, 110, 115, 117, 118, 119, 120, 121, 122, 123, 127, 128, 134, 135, 136, 137, 161, 164, 165
equipment, 108, 128, 149, 161
erosion, 126, 130, 131
ester, 75, 76, 79, 81, 84, 88, 90, 92, 93, 95, 97
ethanol, 61, 66
ethers, 85, 90
ethylene, 61, 66, 71, 72, 81
ethylene glycol, 61, 66, 71, 72, 81
ethylene oxide, 71
Europe, 115
evacuation, 129
evidence, 23, 34
exclusion, 2, 11, 27
exercise, 19, 32
expertise, 36
exposure, 43, 49, 67, 70, 119, 127, 128, 133, 134, 136, 138
extraction, xiii, 42, 117, 118, 119, 120, 121, 123, 125, 126, 127, 128, 129, 130, 132, 136, 155, 166

F

families, 104, 109, 118
federal government, 35, 36, 53, 154, 163, 165, 166
Federal government, 105
federal law, 35, 165
federal mandate, 121, 154
federal regulations, 32, 44, 45
ferrite, 96
fiber, 85, 89

financial, 15, 16, 54, 150
fires, 128
fish, 127, 133, 138, 140
flexibility, 19, 32, 153, 170
fluid, 4, 7, 15, 16, 17, 23, 24, 30, 37, 39, 46, 47, 53, 62, 64, 65, 109, 115, 132, 140, 161, 165, 174, 177
force, 13
formaldehyde, 71, 72, 76, 90
formamide, 68, 82
formation, xi, xii, xiii, 1, 4, 6, 7, 32, 39, 53, 62, 128, 131, 133, 135, 147, 148, 149, 152, 160, 166
fractures, xi, xii, 1, 4, 7, 16, 33, 46, 47, 62, 66, 161
freshwater, 135, 165
friction, 47
funding, 40, 41, 42, 150, 163, 169
funds, 40, 41

G

Garrett County, xiii, 160
gas extraction, xiii, 117, 118, 119, 121, 125, 126, 127, 128, 129, 130, 132, 136, 155
geography, 151
geology, 36, 52, 154
gill, 133
glycol, 61, 66, 68, 71, 72, 78, 81, 82, 84, 86, 92, 96, 97, 128, 136
governments, 35, 146, 175
grants, 12, 40
greenhouse, 119
Ground Water Protection Council (GWPC), 8, 34
groundwater, xii, 3, 7, 8, 9, 10, 13, 30, 32, 34, 36, 37, 38, 39, 44, 45, 46, 48, 52, 54, 55, 105, 113, 114, 115, 118, 131, 132, 168, 174, 176
groundwater aquifer, xii, 115
growth, 4, 7, 24, 116, 144, 150, 151, 154
guidance, 18, 31, 51, 121, 165, 169
guidelines, 39, 121, 134, 140, 141, 177

H

habitat, 162
hazardous wastes, 14
hazards, 34, 65
health, xi, xiii, 23, 25, 30, 33, 39, 44, 49, 50, 55, 60, 61, 62, 63, 64, 65, 68, 70, 71, 72, 73, 103, 104, 105, 108, 110, 111, 117, 118, 119, 120, 122, 123, 124, 125, 126, 127, 128, 130, 133, 134, 136, 139, 141, 148, 155, 160, 162, 163, 164, 165, 173, 175, 176, 177, 178
health and environmental effects, xiv, 160
Health and Human Services, 70, 100
health care professionals, 175
health effects, 39, 71, 128
health problems, 72, 127
heavy metals, 131
height, 53
heptane, 89
HHS, 100
high blood pressure, 72
history, xiv, 44, 103, 144, 155, 173
homeowners, 29, 45
homes, 129
House, vii, 28, 49, 51, 55, 59, 120, 177
House of Representatives, vii, 28, 49, 55, 59, 177
human, 13, 39, 43, 60, 61, 63, 65, 67, 69, 70, 73, 100, 104, 108, 111, 118, 119, 123, 124, 127, 133, 134, 138, 139, 141, 148, 149, 162, 163
human exposure, 43, 119
human health, 39, 60, 61, 63, 65, 68, 70, 73, 104, 108, 111, 118, 123, 124, 141, 148, 162, 163
Hydraulic fracturing, iv, xi, xii, 1, 3, 4, 36, 39, 43, 46, 56, 57, 60, 62, 67, 73, 131, 168
hydrocarbons, 14, 37, 140
hydrogen, 71, 72, 78, 94
hydrogen chloride, 72
hydrogen fluoride, 71
hydroxide, 66, 78, 80, 87, 92, 94

I

identification, 29, 153
identity, 64, 65
improvements, 177
income, 118
independence, xii, 103, 107
Independent Petroleum Association of America (IPAA), 5
individuals, 25
industries, 144
industry, xiv, 4, 7, 9, 19, 21, 32, 33, 34, 36, 41, 45, 57, 62, 64, 104, 105, 108, 111, 119, 122, 123, 144, 145, 146, 147, 150, 151, 154, 155, 156, 157, 160, 168, 169, 170, 171, 177, 178
inflation, 40
infrastructure, 48, 126, 129, 144, 147, 149, 153
ingestion, 67, 133, 138
ingredients, 65
initiation, 21
injections, 11, 30, 63, 64, 110
inspectors, 41, 168
integrity, 16, 17, 149, 161, 162
interest groups, 57
internalizing, 44
intervention, 171
iron, 87
isolation, 33, 108
isomers, 89
issues, xi, xiii, 3, 8, 9, 35, 45, 48, 57, 104, 105, 110, 119, 121, 145, 152, 154, 160, 161, 164, 170, 175

J

Jackson, Lisa, 100
job creation, 144
jurisdiction, xiv, 34, 39, 168

K

ketones, 78

kidneys, 70
kill, 63

L

lakes, 8, 170, 171
landscape, 146, 147, 163
laws, 35, 36, 48, 55, 120, 146, 155
lead, 18, 43, 61, 63, 72, 100, 119, 131, 145, 165, 176
leadership, 104, 146
LEAF, 20, 21, 22, 23, 24, 27, 28, 51, 52
legislation, 2, 3, 10, 24, 32, 46, 49, 146, 162, 163
lifetime, 144
light, 66, 86, 126
liquids, 10, 14, 131, 162
litigation, 27, 33
liver, 67, 70
local conditions, 178
local government, 35, 146, 175, 178
logging, 54
Louisiana, 19, 57, 67, 69, 116, 169, 177

M

magnesium, 134, 137
magnitude, 122, 129
majority, 70
man, 46, 171
management, 8, 30, 34, 37, 38, 43, 48, 55, 57, 121, 146, 148, 149, 150, 151, 152, 154, 156, 157, 162, 163
manganese, 137
mapping, 33
Marcellus shale, xii, 104, 114, 139
Marcellus wells, xii, 107, 130, 160
marketing, 33
marrow, 67
Maryland, viii, xiii, 104, 108, 128, 139, 159, 160, 161, 162, 163, 164, 165, 166
mass, 134, 136
materials, 14, 38, 39, 45, 46, 131, 133, 161, 163

matrix, 39
matter, 22, 135, 151
media, 155, 156
medical, 29, 53, 175
metabisulfite, 94
metals, 104, 109, 118, 131, 132
methanol, 61, 66, 71
Mexico, 19, 25, 67, 69, 70
migration, 38, 63, 115, 131, 149, 161
milligrams, 13, 134
mission, 39, 48, 55, 174
Missouri, 19
mixing, 43
Montana, 2, 3, 19, 25, 34, 67, 69
moratorium, 57, 151

N

naphthalene, 69
national security, xii, 107, 124, 144
natural gas, 1, 3, 4, 5, 6, 7, 11, 12, 17, 22, 27, 28, 30, 34, 35, 36, 42, 44, 49, 60, 62, 63, 73, 98, 99, 103, 104, 105, 107, 108, 109, 111, 116, 117, 118, 119, 122, 124, 125, 126, 128, 130, 131, 143, 144, 145, 147, 148, 149, 152, 155, 166, 167, 168, 170, 173, 174
natural resources, 149, 150, 155, 162, 178
nerve, 72
nervous system, 70
nitrite, 94
North America, 47, 55

O

Obama, 115, 148, 164
Obama Administration, 115
Obama, President, 148, 164
officials, 50, 53, 98, 119, 168, 175
oil production, 1, 5, 173
Oklahoma, 19, 57, 67, 69, 70, 113, 116, 167, 168, 169, 170, 171, 175, 177
Oklahoma Corporation Commission (OCC), xiv, 167

on-shore gas deposits, 143
operations, 2, 9, 10, 11, 18, 20, 23, 27, 28, 30, 32, 34, 36, 37, 39, 44, 52, 63, 64, 108, 110, 113, 118, 122, 123, 126, 127, 128, 132, 145, 149, 150, 151, 162, 164, 165, 168, 169, 170, 176
opportunities, 145, 155
organ, 81
organic chemicals, 118, 132
organic compounds, 53, 131
organic matter, 135
organism, 134, 136
outreach, 120
oversight, 2, 3, 11, 40, 41, 44, 104, 106, 123, 145, 154, 164, 168, 175
oxygen, 128
ozone, 128, 162

P

parallel, 16
pathways, 131, 133, 138
permeability, xi, 1, 3, 4, 5, 6, 46, 174
permit, 12, 17, 24, 33, 41, 56, 57, 58, 109, 110, 127, 135, 136, 139, 153, 160, 162, 163, 165
peroxide, 80, 86, 88
petroleum, 35, 54, 66, 85, 86, 87, 88, 93, 95, 122, 140, 170
Petroleum, 5, 47, 52, 55, 90, 169
petroleum distillates, 54, 66, 86, 87, 95
pH, 134
phenol, 76, 83, 84, 85, 89, 90
phosphate, 76, 79, 84, 91, 93, 94, 97
phosphates, 90
phthalates, 140
plants, 104, 109, 122, 127, 139, 140
playing, 110
poison, 71
policy, 47, 56, 125, 126, 127, 146
pollutants, 48, 61, 68, 71, 72, 73, 104, 110, 118, 121, 127, 128, 176
pollution, 36, 39, 44, 104, 105, 107, 126, 130, 156, 162, 168
polyacrylamide, 78
polyamine, 90
polydimethylsiloxane, 81
polymer, 74, 78, 83, 84, 85, 90, 95, 96
polymers, 17, 75
polypropylene, 92
polyvinylacetate, 92
population, 126, 129
population density, 126, 129
portfolio, 144, 146, 153
potassium, 74, 78, 79
potential benefits, 44
preparation, 47, 163
preparedness, 126
President, 48, 98, 143, 148, 164
prevention, 9, 44, 49, 161, 168
primacy, 11, 18, 19, 32, 41, 42, 50, 51
principles, 58
probability, 129
producers, 30, 64
production costs, 23
production well, 1, 2, 3, 7, 9, 16, 17, 22, 46, 50, 57, 163
professionals, 29, 53, 174, 175, 177
profitability, 144
program administration, 32
project, 15, 42
propylene, 71
prosperity, 146
protection, 10, 11, 25, 30, 32, 34, 36, 37, 44, 45, 46, 48, 50, 54, 55, 57, 119, 134, 150, 164, 165, 166, 168, 170, 175, 176
public concern, 34, 124, 178
public health, xi, xiii, 25, 30, 33, 44, 49, 55, 60, 64, 70, 103, 104, 105, 117, 118, 119, 120, 122, 123, 124, 125, 126, 127, 130, 133, 139, 160, 164, 165, 173, 176, 178
public interest, 57, 58
public safety, 162
publicly owned treatment works (POTWs), 120
pyrolysis, 128
pyrophosphate, 94

Q

quality assurance, 58
quality standards, 152, 154
quartz, 66, 81
quaternary ammonium, 76, 85

R

radiation, 104
radioactive waste, 15
radium, 140
radius, 17
reactions, 135
reality, 115
reasoning, 27
receptors, 133
recession, 144
recognition, 146
recommendations, iv, 38, 48, 146, 148, 156, 165
recovery, 9, 11, 12, 16, 17, 22, 31, 45, 46, 49, 66, 122
recycling, 39
red blood cells, 67
redundancy, 19
reform, 55, 64, 149, 155
Registry, 100
regulations, 9, 10, 11, 12, 13, 14, 19, 20, 21, 22, 23, 24, 27, 29, 30, 31, 32, 34, 35, 36, 37, 38, 39, 40, 41, 44, 45, 50, 51, 52, 53, 54, 55, 56, 57, 65, 71, 99, 110, 121, 122, 149, 151, 152, 154, 155, 175, 176, 177, 178
regulatory agencies, 40, 58, 108
regulatory changes, 146
regulatory framework, 9, 10, 31, 149, 152, 155
regulatory oversight, 145
regulatory requirements, 10, 15, 16, 18, 22, 35, 37, 41, 45
relief, 123
remediation, 15, 154
renewable energy, xiii, 117
repair, 169
requirements, 9, 11, 12, 13, 14, 15, 16, 17, 18, 19, 21, 22, 23, 27, 29, 30, 31, 32, 33, 35, 37, 38, 41, 45, 48, 49, 50, 51, 53, 54, 55, 56, 57, 64, 99, 109, 121, 123, 147, 152, 161, 164, 170, 176
RES, 56
researchers, 42
reserves, 2, 3, 47, 62, 73, 103, 107, 117, 160, 174
residential neighborhood, 126
resolution, 45, 56
resources, 1, 2, 3, 5, 6, 7, 8, 9, 14, 23, 25, 26, 30, 32, 34, 35, 36, 37, 38, 40, 41, 42, 43, 44, 45, 46, 48, 55, 56, 60, 62, 103, 114, 119, 125, 126, 130, 144, 146, 148, 149, 150, 155, 162, 164, 165, 170, 171, 173, 174, 176, 178
response, 23, 24, 36, 37, 46, 54, 56, 110, 121, 129, 149, 151, 162, 177
restoration, 130
restrictions, 110
restructuring, 151
revenue, 41, 122, 144
rights, 35, 160
risk, 2, 7, 34, 38, 39, 41, 63, 65, 119, 126, 127, 129, 134, 136, 138, 153, 161, 163
river systems, 139
rubber, 81
rules, 3, 10, 23, 29, 36, 37, 56, 168, 169, 170, 171, 175, 176, 178
runoff, 164, 170

S

safety, 4, 29, 63, 65, 73, 104, 118, 122, 125, 155, 162, 165
salinity, 131, 165
salts, 76, 79, 83, 90, 92, 93, 95, 104, 109, 118, 131
Saudi Arabia, 107
school, 126, 129
science, 43, 58, 119, 123
scientific understanding, 46

scope, 13, 26, 31, 40, 48, 110, 132, 145, 164, 166
security, 107, 116, 124, 144
sedimentation, 126, 130, 131, 162
sediments, 133, 140
Senate, 28, 167
services, 26, 33, 98, 162
sewage, 104, 127, 132
sheep, 66
showing, 21, 138
signs, 140
silica, 66, 78, 81, 88, 93
silicon, 66
simulation, 33
SiO2, 66, 81
skin, 139
sodium, 65, 74, 78, 79, 80, 81, 82, 87, 91, 92, 95, 96, 98
solid waste, 104, 141
solution, 15, 45, 76, 78, 80, 88, 132
sorption, 93
South Dakota, 19, 114
species, 133, 162, 165
spending, 129
spot market, 155
staffing, 40, 45, 150
stakeholders, 119, 145, 146, 155, 170
state oversight, 40, 175
state regulators, 8, 30, 105, 146, 177
states, 3, 4, 9, 10, 11, 12, 13, 14, 18, 19, 20, 21, 30, 32, 33, 34, 35, 36, 37, 38, 39, 40, 41, 42, 43, 44, 45, 46, 49, 50, 51, 56, 57, 58, 71, 99, 105, 110, 114, 115, 116, 121, 141, 154, 161, 164, 165, 175, 177, 178
statutes, 12, 29, 56, 120, 123
statutory authority, 156
steel, 175
storage, 8, 11, 14, 17, 27, 28, 63, 99, 122
stormwater, 131, 164
stress, 33, 118, 133, 178
strontium, 133, 134, 138
structure, 16, 154
style, 47
sulfate, 77, 78, 79, 81, 83, 85, 89, 93, 94, 95, 96, 98, 134, 138

sulfur, 90
Superfund, 8
suppliers, 62, 72
surfactant, 67, 84, 89
sustainability, 162
sustainable development, 145
sustainable energy, 146

T

tactics, 33
talc, 88
tanks, 39, 63, 132
target, 33
target zone, 33
taxes, 160
teams, 40, 57
technical assistance, 123, 165
technical support, 164
techniques, 14, 46
technologies, 42, 44, 122
technology, 5, 60, 119, 149, 155, 174
terpenes, 80
territory, 50
testing, 8, 9, 48, 139, 161
threats, 25, 104
threshold level, 54
time constraints, 128
tissue, 71
titanate, 90, 97
titanium, 88
toluene, 53, 61, 70, 99
total dissolved solids (TDS), 134, 137, 149
toxic metals, 104
toxic substances, 104
toxicity, 87, 161
trade, 27, 53, 54, 56, 62, 65, 72, 73, 100, 115
transformation, 6
transmission, 122, 168
transparency, 178
transport, 62, 63, 126, 130, 133, 161
transportation, 149
Treasury, 115

treatment, 3, 8, 30, 33, 34, 37, 43, 46, 47, 48, 56, 63, 104, 109, 120, 121, 127, 132, 133, 135, 139, 141, 161
tricarboxylic acid, 74

U

unemployment rate, 144
United, 4, 5, 6, 14, 40, 47, 52, 55, 59, 60, 62, 64, 73, 104, 107, 122, 134, 164, 168, 169
United States, 4, 5, 6, 14, 40, 47, 52, 55, 59, 60, 62, 64, 73, 104, 107, 122, 134, 164, 168, 169
updating, 175
uranium, 15
urethane, 81
USA, 126, 129

V

Vice President, viii, 98, 143
viscosity, 66, 99
volatile organic compounds, 53

W

Washington, 52, 115, 129
waste, 37, 39, 50, 56, 57, 99, 104, 120, 121, 126, 130, 132, 140, 141, 164
waste disposal, 50, 164
waste management, 57
waste treatment, 121
wastewater, 14, 63, 64, 104, 109, 121, 132, 133, 137, 139, 141, 149, 161
water quality, xiii, 3, 8, 25, 39, 43, 64, 118, 124, 125, 126, 130, 148, 152, 154, 165, 168, 174
water quality standards, 152, 154
water resources, 23, 25, 26, 34, 35, 36, 37, 38, 42, 43, 44, 46, 119, 148, 164, 170, 174, 176
water supplies, 7, 24, 45, 60, 63, 105, 139, 149, 168, 176, 178
watershed, 132, 138, 139, 166, 170
waterways, 104, 131, 144, 155
wealth, 177
websites, 138
welfare, 49
wetlands, 162
wildlife, 174
withdrawal, 21, 34, 118, 151
witnesses, 106, 175
workload, 40
World Health Organization, 100

Z

zirconium, 76, 95, 97, 98